PROBABILISTIC AND STATISTICAL METHODS IN COMPUTER SCIENCE

Probabilistic and Statistical Methods in Computer Science

by

Jean-François Mari

LORIA and Université Nancy 2, France

and

René Schott

Université Henri Poincaré-Nancy 1, France

KLUWER ACADEMIC PUBLISHERS

BOSTON / DORDRECHT / LONDON

A C.I.P. Catalogue record for this book is available from the Library of Congress.

Published by Kluwer Academic Publishers,
P.O. Box 17, 3300 AA Dordrecht, The Netherlands.

Sold and distributed in North, Central and South America
by Kluwer Academic Publishers,
101 Philip Drive, Norwell, MA 02061, U.S.A.

In all other countries, sold and distributed
by Kluwer Academic Publishers,
P.O. Box 322, 3300 AH Dordrecht, The Netherlands.

Printed on acid-free paper

ISBN 978-1-4419-4877-9

to our children: Anne,
Cécile, Léon, Marjorie,
Mélanie

Contents

List of Figures   ix
List of Algorithms   xi
List of Tables   xiii
Preface   xv

Part I PRELIMINARIES

1. PROBABILISTIC TOOLS   3
  1.   Random variables, moments, Bayes formula   3
  2.   Law of large numbers   9
  3.   Probability generating function   9
  4.   Markov chains   10
  5.   Martingales   19
  6.   Real-valued random variables   19
  7.   Central limit theorem   24
  8.   Large deviations   25
  9.   Stochastic processes   26

2. STATISTICAL TOOLS   29
  1.   Tests   29
  2.   Distance between two distributions   32
  3.   The EM algorithm for HMM   33
  4.   Unsupervised classification   39

Part II APPLICATIONS

3. SOME APPLICATIONS IN ALGORITHMICS   57
  1.   Probabilistic analysis of Quicksort   57
  2.   Probabilistic analysis of dynamic algorithms 61
  3.   Probabilistic analysis of binary search trees 94
  4.   Probabilistic analysis of genetic algorithms 96

    5.   A markovian concurrency measure   104
    6.   Probabilistic analysis of bin packing problems   111
    7.   Probabilistic analysis of some distributed algorithms   112
    8.   Regular approximation of shuffle products of Context-free Languages   135
    9.   Design and analysis of randomized algorithms   144

4. SOME APPLICATIONS IN SPEECH RECOGNITION   153
    1.   Introduction   153
    2.   Acoustic-front end processor   154
    3.   HMM speech modeling   155
    4.   Second-order HMMs   157
    5.   The extended Viterbi algorithm   160
    6.   The extended Baum-Welch algorithm   161
    7.   Implementation and complexity   165
    8.   Duration model   168
    9.   Experiments on connected digits   169
    10.   Experiments on spelled names over the telephone   171
    11.   Experiments on continuous speech   175

5. SOME APPLICATIONS IN ROBOTICS   177
    1.   Application of hidden Markov models   177
    2.   Application of Markov decision processes   189

Appendices   205
A– Some useful statistical programs   205
    1.   The Gaussian density class   205
    2.   The Centroid class   214
    3.   The Top down clustering program   217

References   221

Index   235

List of Figures

1.1	Graph of the Markov chain	11
1.2	The squadron's life as a Markov chain.	13
1.3	A simple first order HMM	15
1.4	The computation of δ.	16
1.5	Decreasing the order of a HMM_2.	17
2.1	The density of χ_5^2.	32
2.2	Two distributions of paths between phonemes and observations in a sentence.	35
2.3	Distances between groups	45
2.4	Classification and hierarchy.	46
2.5	Input data from a 4 class random generator.	49
2.6	Top down Classification	51
2.7	Final clustering in the 3 class problem.	51
3.1	Binary search trees of size 3	62
3.2	Binary search trees of size 2	62
3.3	A schema	69
3.4	Crossover	97
3.5	Flowchart automaton 1	107
3.6	Flowchart automaton 2	108
3.7	Flowchart automaton 3	109
3.8	Flowchart automaton 4	110
3.9	Evolution of two stacks	113
3.10	New coordinate system	115
3.11	Banker algorithm	118
3.12	Convex hull	147
3.13	Voronoi diagram	149
3.14	Arrangement of line segments	150
4.1	Each state captures a speech segment.	156
4.2	Distributions of duration.	157

4.3	The speech waveform of phonemes [s ɪ].	158
4.4	Typical spectrogram of phonemes: s ɪ.	158
4.5	Variations of $\log(\delta_t(j))$ along the best path in the Viterbi algorithm.	159
4.6	$\log(\delta_t(j))$ assuming model C.	159
4.7	$\log(\delta_t(j))$ assuming the model T.	159
4.8	The best path given by the Viterbi algorithm assuming HMM T.	160
4.9	Example of a time warping alignment.	162
4.10	*a posteriori* transition probability	164
5.1	The Nomad200 robot.	179
5.2	What the robot listens.	180
5.3	Topology of states used for each model of place	180
5.4	The places to learn	181
5.5	The corridor used to make our learning corpus	182
5.6	The segmentation corresponding to a T-intersection	183
5.7	he three sonars used for the segmentation.	183
5.8	The 10 models to recognize.	187
5.9	MDP-based navigation architecture.	193
5.10	The robot navigating to the goal.	195
5.11	An optimal policy.	196
5.12	State aggregation in a corridor.	197
5.13	The policy obtained with the method.	200
5.14	2 missions executed using (a,c) the optimal policy ; (b,d) an abstract policy.	200
5.15	Test environment (3976 states).	201

List of Algorithms

2.1   The Ward's algorithm.   48
2.2   The K-means algorithm   49
4.1   The process of phonetic state tying on OGI.   174
5.1   Transformation of a policy.   199

List of Tables

3.1   Possibility functions in the markovian model.   70

3.2   Possibility functions in Knuth's model   70

4.1   The French phonemes.   161

4.2   String error rates (without post-processing)   170

4.3   String error rates (with post-processing)   170

4.4   Comparison between HMM_1 (with post-processing) and HMM_2 (without post-processing)   170

4.5   Recognition accuracy (in %) for HMM_1 and HMM_2.   172

4.6   Phone recognition rate on $TIMIT$ using HMM 176

5.1   Aggregation benefit.   201

5.2   Results for the test environment.   202

5.3   Policy Execution.   203

Preface

Probability theory and statistics are becoming more and more important in computer science. Almost all fields involve these mathematical tools and it will be difficult for computer scientists to ignore these tools in the future. This book presents a large variety of applications of probability theory and statistics in computer science but we do not cover all domains of application. We focus mostly on:

- Probabilistic algorithm analysis: starting with the Quicksort algorithm, we switch to A.T. Jonassen and D.E.Knuth's famous trivial dynamic algorithm whose analysis isn't. Then we lay out the basics of the probabilistic analysis of dynamic data structures. We illustrate with G. Louchard's method applied to the linear list, then present R. Maier's powerful large deviation analysis. Binary search trees have been deeply investigated over the last decade. We present a nice result on the width obtained by B. Chauvin, M. Drmota and J. Jabbour-Hattab through large deviations. Then we give some hints on R. Cerf's probabilistic analysis of a simple genetic algorithm. A concurrency measure based on simple Markov chain features is defined and analyzed. We present also W.T. Rhee and M. Talagrand's analysis of bin packing problems. Some simple parallel algorithms (two stacks problem, banker algorithms) are analyzed with the same tools as dynamic data structures. Computational geometry (and many other algorithmic fields) benefited largely from probability theory through the design and analysis of randomized algorithms. We restrict ourselves to only a few classical geometric constructions (convex hull, Voronoi diagrams and Delaunay triangulations) and follow an expository survey of O. Devillers.

- Applications of probabilistic tools (Hidden Markov Models, Markov Decision Processes) in speech recognition and robotics. We have decided to work out completely only a few examples. These applications show the important contributions of probabilistic techniques.

This book is written on a self-contained basis: all probabilistic and statistical tools needed are introduced on a comprehensible level. Statistical methods are mostly presented for completeness. They are useful in all fields of Computer Science: statistical algorithm analysis, performance analysis, tests and hypothesis checking, ...
Most of the material is scattered throughout available literature, however, we have nowhere found all of this material collected in accessible form. It is intended for students in computer science, for engineers and for researchers interested in applications of probability theory and statistics. There is didactical material and advanced technical sections.
This book is an outgrowth of a set of lecture notes we wrote for a course we taught in the Department of Computer Science at the University Henri Poincaré-Nancy 1. Probabilistic algorithm analysis started in the 80's. Among the prominent contributors, we want to mention L. Devroye, G. Louchard, R. Maier and B. Pittel. The use of probabilistic and statistical methods in robotics and speech recognition goes back to the 70's.

Most of the applications presented in Part 2 have been developed by the authors in collaboration with colleagues from Loria, Inria-Lorraine, Inria-Sophia Antipolis, Université Libre de Bruxelles, Université de Paris 11-Orsay, Université d'Amiens, Université de Poitiers, University of Arizona, and University of Southern Illinois at Carbondale.
In particular we thank O. Devillers, R. Cerf, D. Geniet, G. Louchard, M. Régnier, W.T. Rhee, M. Talagrand for allowing us to use part of their results or material.

The authors are particularly grateful to O. Aycard, F. Charpillet, D. Fohr, D. Geniet, P. Feinsilver, J.-P. Haton, P. Laroche, G. Louchard, R. Maier, A. Napoli and, L. Thimonier for long term research collaboration. Many examples presented in this book have been developed jointly with the above cited colleagues. J. Atlan, J.M. Drouet and P. Feinsilver helped us with the intricacies of the English language.

I
PRELIMINARIES

Chapter 1

PROBABILISTIC TOOLS

In this chapter, we provide background material on probability theory.

1. RANDOM VARIABLES, MOMENTS, BAYES FORMULA

Consider, for example, a die. Each of the six faces represents a possibility .

DEFINITION 1.1 *An event is a set of possibilities.*

EXAMPLE 1 $A = \{2, 4, 6\}$ *represents the event "An even number appears when playing with a die".*

Notations are borrowed from the theory of sets and we will say that:
The event $A \cap B$ is realized if and only if the event A and the event B are realized.
The event $A \cup B$ is realized if and only if the event A or the event B is realized.
$\bar{A}$ will stand for the event complementary to the event A.

DEFINITION 1.2 *The set of all possibilities of a random experiment is called the universe and will be denoted Ω. $\mathcal{P}(\Omega)$ stands for the set of all subsets of Ω. The set of all possibilities of a random experiment is called the universe and will be denoted Ω. $\mathcal{P}(\Omega)$ stands for the set of all subsets of Ω.*

The cardinality of Ω can be finite or infinite.

3

1.1   THE FINITE UNIVERSE CASE

DEFINITION 1.3 *Given a finite universe Ω. If to each event A is associated a number $P(A)$ called the probability of the event A satisfying the following axioms:*

- $P(\Omega) = 1,$

- $A \cap B = \emptyset \Rightarrow P(A \cup B) = P(A) + P(B).$

The pair (Ω, P) is a probability space.

PROPERTY 1 *For each event A:*

- $0 \leq P(A) \leq 1$

- $P(\bar{A}) = 1 - P(A)$

and

- $P(\emptyset) = 0$

If A and B are two events then:

- $P(A \cup B) = P(A) + P(B) - P(A \cap B).$

DEFINITION 1.4 *Let $\Omega = \{w_1, \ldots, w_n\}$ be a finite probabilty space and let $p_1 = P(w_1), \ldots, p_n = P(w_n)$ be the associated probabilities. The sequence p_i, $1 \leq i \leq n$ is called a probability distribution.*

If $p_1 = p_2 = \ldots = p_n = \frac{1}{n}$ then the probability distribution is called uniform.

DEFINITION 1.5 *Let Ω be a universe. A family of events A_i, $1 \leq i \leq n$ of Ω such that the events A_i are two by two disjoint and their union is Ω is called a partition.*

In other words, a partition consists of a family of events that are mutually exclusisve and exhaustive.

1.2   INFINITE DENUMERABLE UNIVERSE CASE

Some refinements are necessary in this case.

DEFINITION 1.6 *Let Ω be a set. A subset $\mathcal{T}$ of $P(\Omega)$ is a σ-field if:*

- $\Omega \in \mathcal{T},$

- $A \in \mathcal{T} \Rightarrow \bar{A} \in \mathcal{T}$,

- $A_n \in \mathcal{T}, n \in N \Rightarrow \cup_{n=0}^{\infty} \in \mathcal{T}$.

EXAMPLE 2

- $\mathcal{P}(\Omega)$ *is a σ-field*,

- $\{\emptyset, \Omega\}$ *is a σ-field*,

- *Let $A \in P(\Omega)$, then $\{\emptyset, \Omega, A, \bar{A}\}$ is a σ-field called σ-field generated by A.*

In general,

DEFINITION 1.7 *A probability space is a triple $(\Omega, \mathcal{T}, P)$ where:*

- Ω *is a set*,

- $\mathcal{T}$ *is a σ-field on Ω*,

- P *is a probability measure on $\mathcal{T}$:*

$$P : \mathcal{T} \to [0, 1]$$

such that:

- $P(\Omega) = 1$,
  The elements of $\mathcal{T}$ are called events and for $A \in \mathcal{T}$, $P(A)$ is the probability of the event A.

-

$$P(\cup_{n=0}^{\infty} A_n) = \sum_{n=0}^{\infty} P(A_n)$$

for each sequence of events which are pairwise disjoint.

1.3   CONDITIONAL PROBABILITY, BAYES FORMULA

DEFINITION 1.8 *Let $(\Omega, \mathcal{T}, P)$ be a probability space and $B \in \mathcal{T}$ such that $P(B) \geq 0$.*
For each $A \in \mathcal{T}$, the number $P(A|B) = \frac{P(A \cap B)}{P(B)}$ is called the conditional probability of A given B.

DEFINITION 1.9 *Let A_i, $i \in \mathcal{N}$ be a sequence of events of $\mathcal{T}$. A_i, $i \in N$ is a complete system of events if:*

- *The A_i are pairwise disjoint,*

- $P(\cup_i A_i) = 1.$

THEOREM 1.1 *Let A_i, $i \in \mathcal{N}$, be a complete system of events and $A \in \mathcal{T}$. Then*

$$P(A) = \sum_i P(A|A_i)P(A_i).$$

Bayes' formula follows from the above definition and theorem:

COROLLARY 1 *(Bayes' formula)*
Assume that the numbers $P(A|A_i)$, $P(A_i)$, $i \in \mathcal{N}$ are known. Then

$$P(A_i|A) = \frac{P(A|A_i)P(A_i)}{\sum_i P(A|A_i)P(A_i)}.$$

REMARK 1 *Bayes formula is used for prediction and has numerous applications in computer science as we will see later.*

DEFINITION 1.10 *Two events A and B are independent for the probability P if:*

$$P(A \cap B) = P(A)P(B)$$

Three events A, B, C are mutually independent if they are pairwise independent and if:

$$P(A \cap B \cap C) = P(A)P(B)P(C)$$

More generally, $A_1, \ldots, A_n$ are mutually independent if they are k-by-k independent for $k = 2, \ldots, n$.

1.4   DISCRETE RANDOM VARIABLES

DEFINITION 1.11 *Let $(\Omega, \mathcal{T}, P)$ be a probability space. A random variable X is a mapping*

$$X : \Omega \to \mathcal{R}$$

Let X be a random variable taking the values $x_1, \ldots, x_n, \ldots$. The sequence of numbers $p_i = P(X = x_i)$ is called the distribution (or law) of the random variable X.

EXAMPLE 3 *Here are some classical discrete random variables.*

a) Bernoulli random variable
Let $p \in [0,1]$, the random variable X given by:

$$P(X = 1) = p \text{ and } P(X = 0) = 1 - p$$

is called a Bernoulli random variable (or Bernoulli trial)

b) Binomial random variable
Repeat n times the Bernouilli trial, the random variable Y which counts the number of ones is called an (n,p)-binomial random variable. Its distribution is given by

$$P(Y = k) = C_n^k p^k (1 - p)^{n-k}, \ k \in \mathcal{N}$$

where $C_n^k = \frac{n!}{k!(n-k)!}$

c) Pascal's (or geometric) random variable
Repeat n times the Bernouilli trial. The random variable Z which corresponds to the first appearence of the number 1 is called a Pascal (or geometric) random variable. Its distribution is given by

$$P(Z = k) = (1 - p)^{k-1} p, \ k \in \mathcal{N}$$

d) Poisson random variable
Let λ be a stricly positive real number. The random variable U whose distribution is given by

$$P(U = k) = e^{-\lambda} \frac{\lambda^k}{k!}, \ k \in \mathcal{N}$$

is called a Poisson random variable with parameter λ.

1.5   MOMENTS OF A RANDOM VARIABLE

DEFINITION 1.12 *Let X be a discrete random variable whose distribution is $p_i = P(X = x_i), \ldots, p_n = P(X = x_n), \ldots$ If the series $\sum_i p_i x_i$ is absolutely convergent (e.g. $\sum_i p_i |x_i| < \infty$), then $E(X) = \sum_i p_i x_i$ is called the expected value (or mean, or first moment) of X.*
If $E(X^r) = \sum_i p_i x_i^r < \infty$, this number is the rth moment of X.
If $E((X - E(X))^2) < \infty$, this number is called the variance of X, $\mathrm{var} X$, and $\sigma_X = \sqrt{\mathrm{var} X}$ is the standard deviation of X.
A random variable X such that $E(X) = 0$ and $\sigma_X = 1$ is called centered and scaled random variable.

EXERCICE 1 *Show that the (n,p)-binomial random variable X has expected value $E(X) = np$.*
Show that the Poisson random variable Y of parameter λ has expected value $E(Y) = \lambda$.

Below we collect several classical properties:

PROPERTY 2 $E(aX + b) = aE(X) + b$, $\forall (a,b) \in \mathcal{R}^2$,
$var X = E(X^2) - (E(X))^2$,
$var(aX) = a^2 var(X)$, $\forall a \in \mathcal{R}$,
If $X_1, \ldots, X_n$ are random variables having first moments then

$$E(X_1 + \ldots + X_n) = E(X_1) + \ldots + E(X_n)$$

If, in addition, $X_1, \ldots, X_n$ are independent random variables then

$$E(X_1 \ldots X_n) = E(X_1) \ldots E(X_n)$$

If $X_1, \ldots, X_n$ are independent random variables having variances then

$$var(X_1 + \ldots + X_n) = var(X_1) + \ldots + var(X_n).$$

1.6 INEQUALITIES

The following inequalities are the most commonly used:

a) Markov inequality
Let X be a positive random variable whose first moment $E(X)$ exists, then for all $\lambda \geq 1$

$$P(X \geq \lambda E(X)) \leq \frac{1}{\lambda}$$

b) Bieynamé-Tchebychev inequality (BT inequality)
Let X be a random variable with expected value m and variance σ_X^2. Then for all real number $a \geq 0$,

$$P(\frac{|X - m|}{\sigma_X} > a) \leq \frac{1}{a^2}$$

c) Chernoff bound
Let $X_1, \ldots, X_n$ be a sequence of independent Bernoulli random variables with the same law $P(X_k = 1) = p$ and $P(X_k = 0) = 1 - p$, for $k \in \mathcal{N}$. Let $S_n = \sum_1^n X_k$, then for all $t \geq 0$,

$$P(|S_n - np| \geq nt) \leq 2e^{-2nt^2}$$

2. LAW OF LARGE NUMBERS

The weak law of large numbers can be stated as follows:

THEOREM 1.2 *Let* $X_1, \ldots, X_n$ *be a sequence of independent random variables with the same law (i.e. i.i.d. random variables). Assume that the mean m and the variance of* X_k *exist. Then for all* $\epsilon > 0$:

$$lim_{n\to\infty} P(|\frac{X_1 + \ldots + X_n}{n} - m| > \epsilon) = 0.$$

This theorem follows from BT's inequality.

The law of large numbers has the following practical interpretation:

Let E be a random experiment and A an event which is realized with probability p. Let $N(A)$ be the number of occurrences of A among N independent repetitions of E. Then for all $\epsilon > 0$:

$$P(|\frac{N(A)}{N} - p| > \epsilon) \to_{N\to\infty} 0$$

The strong law of large numbers tells us that:

THEOREM 1.3 *Under the same assumptions as before, the event*

$$\{w \in \Omega \mid lim_{n\to\infty} \frac{X_1(w) + \ldots + X_n(w)}{n} = p\}$$

has a probability equal to 1.

3. PROBABILITY GENERATING FUNCTION

DEFINITION 1.13 *Let* X *be a random variable with values in* $\mathcal{N}$ *and* $p_n = P(X = n)$. $G_X(x) = \sum_0^\infty p_n x^n$ *is the generating function of* X.

EXAMPLE 4

- *If* X *is a Bernoulli random variable with parameter* p *then* $G_X(x) = (1 - p + px)^n$.

- *If* X *is the Poisson random variable with parameter* $\lambda > 0$, *then* $G_X(x) = e^{\lambda(x-1)}$.

The following result has useful applications:

THEOREM 1.4 *Let* G_X *and* G_Y *be the generating functions of two independent random variables* X *and* Y. *Then the generating function* G_{X+Y} *of the random variable* $X + Y$ *is given by:*

$$G_{X+Y}(x) = G_X(x).G_Y(x).$$

From this theorem we can easily derive the generating function of the (n,p)-binomial random variable Z:

$$G_Z(x) = (1 - p + px)^n$$

In fact, the mean and variance of a random variable can be easily derived from its generating function G_X as follows:

THEOREM 1.5 *Assume that X is a random variable whose mean and variance exist. Then:*

$$E(X) = G'_X(1)$$

and

$$\text{var}\,X = G''_X(1) + G'_X(1) - (G'_X(1))^2.$$

4.   MARKOV CHAINS

DEFINITION 1.14 *Consider a probability space $(\Omega, \mathcal{F}, P)$ and a sequence (X_n), $n \in \mathcal{N}$ of random variables whose values are in a finite or enumerable space E.*
$(X_n), n \in \mathcal{N}$ is a Markov chain if:

$$P(X_{n+1} = j \mid X_0, \dots, X_n) = P(X_{n+1} = j | X_n), \forall j \in E.$$

In other terms, the "future" depends only on the "present".

DEFINITION 1.15 *If the transition probabilities $P(X_{n+1} = j \mid X_n = j)$ do not depend on n, the Markov chain is called (time-) homogeneous and we denote $p(i,j) = P(X_{n+1} = j \mid X_n = i)$.*

The following properties are obvious:

PROPERTY 3

- $p(i,j) \geq 0$

- $\sum_{j \in E} p(i,j) = 1$.

A Markov chain is given either by a transition matrix or by a labelled oriented graph.

EXAMPLE 5

$$\begin{pmatrix} \frac{1}{2} & 0 & \frac{1}{2} & 0 & 0 \\ 0 & \frac{1}{4} & 0 & \frac{3}{4} & 0 \\ 0 & 0 & \frac{1}{3} & 0 & \frac{2}{3} \\ \frac{1}{4} & \frac{1}{2} & 0 & \frac{1}{4} & 0 \\ \frac{1}{3} & 0 & \frac{1}{3} & 0 & \frac{1}{3} \end{pmatrix}$$

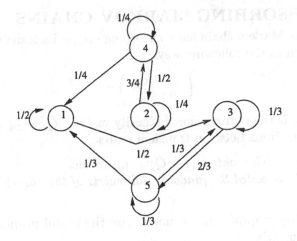

Figure 1.1.   Graph of the Markov chain

The corresponding oriented graph is given by Figure 1.1.

PROPERTY 4 *Let M be the transition matrix of a Markov chain. Then M^n is the transition matrix of an n-step Markov chain.*

4.1   CLASSIFICATION OF THE STATES

DEFINITION 1.16 *A state j is called accessible from a state i if there exits $n > 0$ such that $p_n(i,j) > 0$ (e.g. the probability of reaching i from j in n steps is strictly positive). We denote: $i \rightarrow j$.*
Two states i and j communicate if $i \rightarrow j$ and $j \rightarrow i$.
A set C of states is closed if $p(i,j) = 0$, $\forall i \in C$, $\forall j \notin C$.
If all states communicate, the Markov chain is called irreducible.
If for $i \in E$, $p(i,j) = 0$, $\forall j \neq i$, i is called an absorbing state.
A state $i \in E$ is periodic with period $d(i)$ if:

$$d(i) = GCD(n \geq 1, \ 0 < p_n(i,i) < 1)$$

where GCD means greatest common divisor. If $d(i) = 1$, i is aperiodic. Let q_i be the probability that when leaving the state i, the Markov chain comes back to i:

- *if $q_i = 1$, the state i is called recurrent,*

- *if $q_i < 1$, the state i is called transient,*

- *A state i is positive recurrent if the expected time of return to i is finite,*

- *A state i which is aperiodic and positive recurrent is called ergodic.*

4.2   ABSORBING MARKOV CHAINS

If an n-state Markov chain has r absorbing states, its transition matrix can be written in the following way:

$$\begin{pmatrix} I & (O) \\ R & Q \end{pmatrix}$$

where $I = I_{r,r}$ is the r-dimensional identity matrix. $Q = Q_{n-r,n-r}$ is the matrix of transitions between transient states.

PROPOSITION 1   *The matrix $(I - Q)$ is invertible.*
$N = (I - Q)^{-1}$ *is called the fundamental matrix of the absorbing Markov chain.*

In the following proposition, we summarize the useful properties of the fundamental matrix.

PROPOSITION 2   *Let $N = (n_{ij})$ be the fudamental matrix of an absorbing Markov chain. Then:*

- n_{ij} *is the average number of steps taken by the Markov chain for reaching the state j when starting from the state i,*

- $\sum_{j=1}^{j=n-r} n_{ij}$ *is the average number of steps until absorption when starting from state i.*

Let $A = (a_{ik}) = N.R$, a_{ik} is the probability of being absorbed by the absorbing state k when starting from state i.

4.3   STATIONARITY

DEFINITION 1.17   *A vector $\Pi = (\Pi(i), i \in E)$ such that $\Pi(i) > 0, \forall i \in E$ and $\sum_{i \in E} \Pi(i) = 1$ which satisfies the relation $\Pi = \Pi M$ is called stationary distribution of the Markov chain.*

Criteria for existence and uniqueness of a stationary distribution are given below.

THEOREM 1.6   *Let (X_n), $n \in \mathcal{N}$ be an irreducible, aperiodic Markov chain.*

- *If X_n is positive recurrent then $lim_{n \to \infty} p_n(i,j) = \Pi(j) > 0, \forall i \in E$. $(\Pi(i), i \in E)$ is a probability distribution uniquely defined by $\Pi = \Pi.M$.*

- *If the states of (X_n), $n \in \mathcal{N}$ are transient or null recurrent then*

$$lim_{n \to \infty} p_n(i,j) = 0, \ \forall (i,j) \in E^2$$

and no stationary distribution does exist.

4.4   HIDDEN MARKOV MODEL

We introduce the concept of Hidden Markov model (HMM) through the following example:

EXAMPLE 6 *A four plane squadron performs missions over an enemy country and suffers losses. Let p be the probability that a plane is missing and $q = 1-p$ the probability that a plane comes safety back to the airfield. The mission is launched when the number of planes is strictly greater than 2. Each night, if the number of planes is less than 3, the squadron receives another plane.*

This system can be described by a set of states E_1, E_2, E_3, E_4 associated with the number of planes (1, 2, 3 or 4) available each morning as shown in figure 1.2. The transition probability can be determined as follows:

$E_1 \rightarrow E_2$ when the squadron has only one plane, it receives a second plane during the night with a probability of 1;

$E_2 \rightarrow E_3$ like the transition $E_1 \rightarrow E_2$;

$E_3 \rightarrow E_2$ when one plane is missing among 3, the probability of this transition is $3pq^2$;

$E_4 \rightarrow E_3$ this transition occurs in two cases: either one plane is missing among 4 -- probability $4pq^3$ -- or 2 planes are missing among 4 -- probability $6p^2q^2$ -- and the squadron is provided with one plane during the night.

The other probabilities can be easily derived by applying the binomial law.

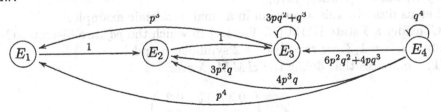

Figure 1.2.   The squadron's life as a Markov chain.

After a mission, the pilots always have drinks in the squadron's bar where the atmosphere is highly correlated with the difficulties that they encountered during the last flight. Now, we suppose that the airfield is forbidden to the public. Therefore, the only way to know what state the squadron is in, is to analyze the atmosphere in the bar every day. This leads to the definition of a hidden Markov process where the observations are drawn from the bar and where the squadron's states are hidden from the observer. A classical problem is to retrieve the sequence of states of the squadron in the light of the observations drawn from the bar.

DEFINITION 1.18 *A first-order hidden Markov Model (HMM_1) λ is described by:*

- $\mathbf{S} = \{s_1, s_2, \dots, s_N\}$, *a set of N states,*

- *A transition matrix $\mathbf{A}$ over $S \times S$, where the i th, j th element a_{ij} is the a priori probability of a transition from state s_i at time t to state s_j at time $t+1$. Due to the Markov assumptions this matrix is time independent,*

- $b_i(.)$ *the probability density functions (pdf) associated with the states s_i.*

Without loss of generality, we will assume that the initial state is s_1 – the lowest index – and the last state s_N is the final state of the Markov chain.

4.5   THE VITERBI ALGORITHM

The Viterbi algorithm plays an important role in stochastic modeling. Its purpose is to extract the best states sequence that produces the observed sequence. It is an iterative algorithm, linear in t and polynomial in N, depending on the order of the model. Its first description was given by D. Forney [Forney, 1973].

Let us describe this algorithm in a small academic example.

Consider a 3 state HMM (cf. Fig 1.3) in which the *pdf* are discrete. On state i, $i = 1, 3$, we can observe 2 symbols: a and b.

The matrices that define the HMM_1 λ are:

$$A = \begin{pmatrix} 0.3 & 0.5 & 0.2 \\ 0 & 0.3 & 0.7 \\ 0 & 0 & 1 \end{pmatrix}$$

that defines the transition probabilities between the 3 states, and

$$B = \begin{pmatrix} 1 & 0 \\ 0.5 & 0.5 \\ 0 & 1 \end{pmatrix}$$

in which a row represents a *pdf*. The initial state probabilities at time
0 are given by the vector:

$$\pi = \begin{pmatrix} 0.6 \\ 0.4 \\ 0 \end{pmatrix}$$

Given a sequence of T observations $y = (y_1, y_2, \ldots, y_T)$, the algorithm

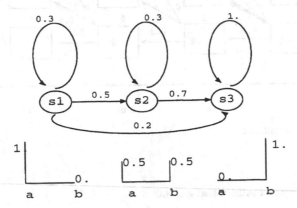

Figure 1.3.   A simple first order HMM

extracts the best path

$$S = (s_1, s_2, \ldots, s_T)$$

by computing the best partial path from $t = 1$ up to time t and ending
in state v for ecah state v. We define $\delta_t(v)$ as the probability of this best
partial path.

$$\delta_t(v) = \max_{s_t}[P(s_1, \ldots, s_{t-2}, s_{t-1}, s_t = v, y_1, \ldots, y_t | \lambda)], \quad 1 \leq v \leq N$$

$$1 \leq t \leq T$$

A recursive computation is given by the following equation:

$$\delta_t(v) = \max_{1 \leq u \leq N}[\delta_{t-1}(u) \cdot a_{uv}] \cdot b_v(y_t), \quad 1 \leq v \leq N \qquad (1.1)$$

$$1 \leq t \leq T$$

The initial condition is given by:

$$\delta_0(i) = \Pi_i, \quad 1 \leq i \leq N$$

Figure 1.4 shows how the algorithm works when the observed sequence
is *aabb*.

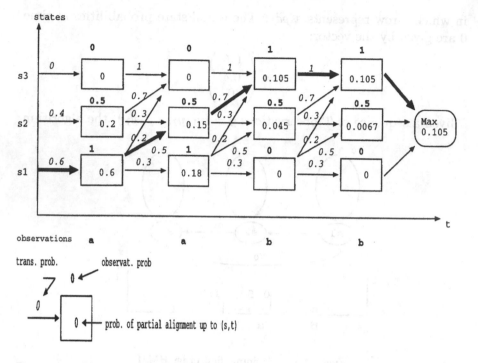

Figure 1.4.   The computation of δ based on the sequence of observations *aabb* (from Kriouile [Kriouile, 1990]).

Second-order HMM (HMM2).   Unlike the first-order Markov chain, where the stochastic process is specified by a 2-dimensional matrix of *a priori* transition probabilities a_{ij} between states s_i and s_j, the second-order Markov chain is specified by a 3 dimensional matrix a_{ijk}.

$$a_{ijk} \stackrel{\text{def}}{=} P(q_t = s_k | q_{t-1} = s_j, q_{t-2} = s_i, q_{t-3} = ...) =$$
$$P(q_t = s_k | q_{t-1} = s_j, q_{t-2} = s_i) \qquad (1.2)$$

with the constraints:

$$\sum_{k=1}^{N} a_{ijk} = 1 \quad \text{with } 1 \leq i \leq N \, , \; 1 \leq j \leq N$$

where N is the number of states in the model and q_t is the actual state at time t.

The probability of the state sequence $Q \stackrel{\text{def}}{=} q_1, q_2, ..., q_T$ is defined as:

$$P(Q) = \Pi_{q_1} a_{q_1 q_2} \prod_{t=3}^{T} a_{q_{t-2} q_{t-1} q_t} \qquad (1.3)$$

where Π_i is the probability of state s_i at time t = 1 and a_{ij} is the probability of the transition $s_i \to s_j$ at time t = 2.

DEFINITION 1.19 *A second-order hidden Markov Model (HMM2)* λ *is described by:*

- $S = \{s_1, s_2, \ldots, s_N\}$, *A set of N states;*

- *A transition matrix* **A** *over S X S x S, where the ith, jth, kth element* a_{ijk} *is the a priori probability of a transition* $s_i \to s_j \to s_k$ *between time t − 1 and t + 1. Due to the Markov assumptions this matrix is time independent.*

- $b_i(.)$ *the probability density functions (pdf) associated with the states* s_i.

Given a sequence of observed vectors $O \overset{\text{def}}{=} O_1, O_2, \ldots, O_T$, the joint state-output probability $P(Q, O|\lambda)$, is defined as:

$$P(Q, O|\lambda) = \Pi_{q_1} b_{q_1}(O_1) a_{q_1 q_2} b_{q_2}(O_2) \prod_{t=3}^{T} a_{q_{t-2} q_{t-1} q_t} b_{q_t}(O_t) \qquad (1.4)$$

Each second-order Markov model has an equivalent first-order model on the 2-fold product space **S x S**, but going back to first-order increases the number of states dramatically. For instance, figure 1.5(b) shows the equivalent HMM_1 associated with the HMM_2 depicted in figure 1.5(a). In this model the states in the same column share the same *pdf*.

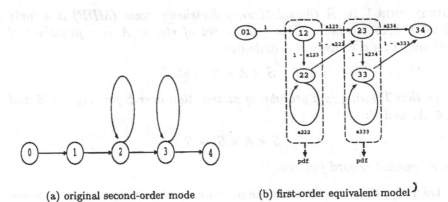

(a) original second-order mode   (b) first-order equivalent model

Figure 1.5.   Decreasing the order of a HMM_2.

This topology has already been investigated in several works [Russell and Cook, 1887, Suaudeau and André-Obrecht, 1993]. It is interesting to note that HMM_2 converge naturally to Ferguson-like models.

The transition probabilities determine the characteristic of the state duration model. In a HMM_1, whose topology is depicted in the figure 1.5(a), the probability $d_j(l)$ that the stochastic chain loops l times in the state j follows a geometric law of parameter a_{jj}.

$$d_j(l) = a_{jj}^{l-1} \times (1 - a_{jj})$$

In the model depicted in figure 1.5(b), in which the successive states are denoted by $i = j - 1$, j, $k = j + 1$, the duration in state j may be defined as :

$$
\begin{aligned}
d_j(0) &= 0 \\
d_j(1) &= a_{ijk}, \quad i \neq j \neq k \\
d_j(n) &= (1 - a_{ijk}) \cdot a_{jjj}^{n-2} \cdot (1 - a_{jjj}), \quad n \geq 2
\end{aligned}
\tag{1.5}
$$

The state duration in a HMM_2 is governed by two parameters, *i.e*, the probability of entering a state only once, and the probability of visiting a state at least twice, with the latter modeled as a geometric decay. This distribution fits a probability density of durations [Crystal and House, 1988] better than the classical exponential distribution of an HMM_1. This property is of great interest when an HMM_2 models a phoneme in which a state captures only 1 or 2 frames. In the framework of speech recognition, an interesting example related to phoneme duration is given on page 157.

4.6   MARKOV DECISION PROCESSES

DEFINITION 1.20 *A (finite) Markov decision process (MDP) is a tuple $< S, A, T, R >$ where: S is a finite set of states; A is a finite set of actions; T is a transition distribution:*

$$T : \; S \times A \times S \to [0, 1]$$

such that $T(s, a, .)$ is a probability distribution over S for any $s \in S$ and $a \in A$; and R:

$$S \times A \times S \to \mathcal{R}$$

is a bounded reward function.

Intuitively, $T(s, a, w)$ denotes the probability of moving to state w when action a is performed at state s, while $R(s, a, w)$ denotes the immediate utility associated with the transition $s \to w$ under action a. We recommend M. Puterman's book [Puterman, 1994] for details.

We will see in Part 2 that given an MDP, the objective is to construct a policy that maximizes expected accumulated reward over some domain H of interest. H will be called the horizon.

4.7   POMDP

Partially observable Markov decision processes (POMDP) have also been defined and applied in domains where information/observation is only partially available.

DEFINITION 1.21 *A POMDP* $M =< S, A, T, R, Z, O >$ *is defined in part by an MDP model: a finite set S of states, and a finite set A of actions, a transition function T, and a reward function R. In addition, it includes a finite set of observations, Z and an observation function*

$$O : S \times A \to Z.$$

The quantity $O(s', a, z)$ is the probability of observing $z \in Z$ in state s' after taking action a.

5.   MARTINGALES

DEFINITION 1.22 *Given a σ-field $\mathcal{A}$ on a set Ω, a real-valued function u on Ω is called $\mathcal{A}$-measurable if for each t the set of all points where $u(t) \leq t$ belongs to $\mathcal{A}$.*

DEFINITION 1.23 *Let $(X_n), n \in \mathcal{N}$ be a sequence of random variables and $(Z_n), n \in \mathcal{N}$ a sequence of X_n -mesurable random variables. $Z_n, n \in \mathcal{N}$ is a martingale if*

$$E(Z_{n+1}|X_1, \ldots, X_n) = Z_n.$$

In other words, the expected value at time $n + 1$ (the future) knowing past and present is equal to the value obtained at time n (present).

THEOREM 1.7 *Let $Z_n, n \in \mathcal{N}$ be a martingale with $E(Z_n^2) < \infty$ for all n. Then there exists a random variable Z such that $Z_n \to Z$ in law and in L^2 (space of square integrable functions).*
Moreover, if $Z_n, n \in \mathcal{N}$ satisfies $E(|Z_n|^p) < C_p$ (constant depending only of p) for some $p > 2$, then

$$Z_n \to Z \in L^p.$$

6.   REAL-VALUED RANDOM VARIABLES

In this section we consider random variables $X : \Omega \to \mathcal{R}$

DEFINITION 1.24 *The distribution function of a random variable X is the function $F_X : \mathcal{R} \to [0, 1]$ given by $F_X(x) = P(X \leq x)$.*

PROPERTY 5 *F_X has the following properties:*

- F_X *is an increasing function,*

- F_X *is a right continuous function (a each point),*

- $lim_{x \to -\infty} F_X(x) = 0$ *and* $lim_{x \to \infty} F_X(x) = 1$.

PROPOSITION 3 *If F is a function satisfying the three properties above then there exists a random variable whose distribution function is F.*

DEFINITION 1.25 *A function* $f : \mathcal{R} \to \mathcal{R}$ *is a (probability) density of the random variable X if:*

$$\forall x \in \mathcal{R}, \ F_X(x) = \int_{-\infty}^{x} f(u)du.$$

This definition generalizes to higher dimensions. Consider a random vector $X = (X_1, \dots, X_n)$. $f : \mathcal{R}^n \to \mathcal{R}$ is a density if for each subset $I \in R^n$,

$$P(X \in I) = \int_I f(x_1, \dots, x_n)dx_1 \dots dx_n.$$

EXAMPLE 7

- $f(x) = \lambda e^{-\lambda x}, \ x > 0, \ \lambda > 0$ *is the density of a random variable called exponential random variable.*

-

$$f(x) = \frac{1}{\sqrt{2\pi}\sigma} e^{(-\frac{1}{2}(\frac{x-m}{\sigma})^2)}$$

   is the density of the gaussian (or normal) random variable X with expected value m and variance σ^2 *usually denoted* $N(m, \sigma)$.

6.1   MARGINAL DISTRIBUTIONS

Assume that the pair of random variables (X_1, X_2) has a density $f(x_1, x_2)$. Then the density of the random variable X_1 is given by

$$f_{X_1}(x_1) = \int_R f(x_1, x_2)dx_2 = f(x_1, .)$$

and the density of X_2 is

$$f_{X_2}(x_1) = \int_R f(x_1, x_2)dx_1 = f(., x_2).$$

These definitions generalize to n-uples.

THEOREM 1.8 *Let $X_1, \ldots, X_n$ be random variables with respective densities $f_{X_1}, \ldots, f_{X_n}$. $X_1, \ldots, X_n$ are independent if and only if*

$$f(x_1, \ldots, x_n) = f_{X_1}(x_1) \ldots f_{X_n}(x_n).$$

6.2 DENSITY OF A SUM OF TWO RANDOM VARIABLES

THEOREM 1.9 *Assume that the random variables X and Y have respective densities f_X and f_Y. If X and Y are independent then $X + Y$ has a density given by:*

$$f_{X+Y}(t) = \int_{-\infty}^{+\infty} f_X(x) f_Y(t-x) dx$$

$$= \int_{-\infty}^{+\infty} f_Y(y) f_X(t-y) dy$$

$$= f_X * f_Y(t).$$

$*$ *is called convolution product.*

6.3 MOMENTS

The quantities are defined as previously with the integral replacing the sum.

DEFINITION 1.26 *Let X be a random variable with density f_X. If the integral $\int_{-\infty}^{+\infty} x f_X(x) dx$ exists then $E(X) = \int_{-\infty}^{+\infty} x f_X(x) dx$ is called the expectation (or mean or first moment) of the random variable X.*
Higher moments are defined similarly.
The variance of X is

$$var(X) = E(X^2) - (E(X))^2 = \int_{-\infty}^{+\infty} x^2 f_X(x) dx - (\int_{-\infty}^{+\infty} x f_X(x) dx)^2.$$

EXAMPLE 8

- Let X be the exponential random variable with density $f(x) = \lambda e^{-\lambda x}$, $\lambda > 0$. Then

$$E(X) = \int_{-\infty}^{+\infty} x \lambda e^{-\lambda x} dx = \frac{1}{\lambda}$$

and

$$var X = \frac{1}{\lambda^2}$$

- As mentioned previously $f(x) = \frac{1}{\sqrt{2\pi}\sigma}e^{(-\frac{1}{2}(\frac{x-m}{\sigma})^2)}$ is the density of the gaussian (or normal) random variable X with expected value m and variance $(\sigma)^2$ usually denoted $N(m, \sigma)$.

6.4   CONDITIONAL RANDOM VARIABLES

DEFINITION 1.27 *Let (X, Y) be a pair of random variables with density $f(x, y)$. For each $x \in R$ such that the marginal distribution $f(x, .) \neq 0$, the function: $y \to f(y|x) = \frac{f(x, y)}{f(x, .)}$ is a probability density called conditional density of Y given $\{X = x\}$. If x varyes, we get a family of densities called conditional densities of Y given X.*

DEFINITION 1.28 *If $E(Y|x) = \int_{-\infty}^{+\infty} y f(y|x) dy$ exists, $E(Y|x)$ is called the conditional expectation of Y given $\{X = x\}$.*
The map of the function

$$l : x \to E(Y|x)$$

is the regression curve of Y with respect to X.

6.5   COVARIANCE

DEFINITION 1.29 *The covariance of the pair (X, Y) of random variables is defined by:*

$$cov(X, Y) = E[(X - E(X))(Y - E(Y))] = E(XY) - E(X)E(Y).$$

REMARK 2 *If X and Y are independent then $E(XY) = E(X)E(Y)$ and $cov(X, Y) = 0$ (but the converse is false!).*

THEOREM 1.10 *Let (X, Y) be a pair of random variables. Then:*

$$var(X + Y) = var(X) + var(Y) + 2cov(X, Y)$$

More generally:

$$var(X_1 + \ldots + X_n) = \sum_{i=1}^{n} var(X_i) + 2 \sum_{j,k;j<k} cov(X_j, X_k).$$

REMARK 3 *If X and Y are independent then $cov(X, Y) = 0$ and $var(X + Y) = var(X) + var(Y)$ as announced previously.*

6.6   CORRELATION COEFFICIENT AND REGRESSION LINE

DEFINITION 1.30 *Let X, Y be two random variables, $(\sigma)_X^2 = var X$, $\sigma_Y^2 = var Y$ and $X^* = \frac{X - E(X)}{\sigma_X}$, $Y^* = \frac{Y - E(Y)}{\sigma_Y}$.*

The correlation coefficient of (X,Y) denoted $\rho(X,Y)$ is defined by:

$$\rho(X,Y) = cov(X^*,Y^*) = \frac{cov(X,Y)}{\sigma_X \sigma_Y}.$$

We summarize below few properties of ρ.

PROPERTY 6

- $-1 \leq \rho(X,Y) \leq 1$,

- $\rho(X,Y) = 1$ *or* -1 *if and only if there exist two constants a and b such that $Y = aX + b$ except on a subset of values of X of probability 0.*

- $\forall\ a_1,\ a_2,\ b_1,\ b_2 \in \mathcal{R}$ *with* $a_1 > 0,\ a_2 > 0$:

$$\rho(a_1 X + b_1, a_2 Y + b_2) = \rho(X,Y)$$

DEFINITION 1.31 *The line (L) whose equation is:*

$$y - E(Y) = \frac{\sigma_Y}{\sigma_X}\rho(X,Y)(x - E(X))$$

is called the regression line of Y with respect to X.

6.7 CHARACTERISTIC FUNCTION

DEFINITION 1.32 *The characteristic function of the random variable X is the complex-valued function ϕ_X defined on R by:*

$$\phi_X(t) = E(e^{itX}) = E(\cos tX) + iE(\sin tX)$$

If X has a density f then:

$$\phi_X(t) = \int_{-\infty}^{+\infty} e^{itx} f(x)dx.$$

PROPERTY 7 *If X and Y are random variables such that $Y = aX + b$, $(a,b) \in R^2$, then:*

$$\phi_Y(t) = e^{ibt}\phi_X(at).$$

EXAMPLE 9 *Let X be a gaussian random variable with mean μ and variance σ^2 (i.e. $X = N(\mu,\sigma)$), then:*

$$\phi_X(t) = e^{i\mu t - \frac{\sigma^2 t^2}{2}}$$

If $X = N(0,1)$, then $\phi_X(t) = e^{-\frac{t^2}{2}}$.

THEOREM 1.11 *If $X_1 \ldots X_n$ are independent random variables with respective characteristic functions $\phi_{X_1}, \ldots, \phi_{X_n}$, then:*

$$\phi_{X_1+\ldots+X_n}(t) = \phi_{X_1}(t) \ldots \phi_{X_n}(t).$$

7.   CENTRAL LIMIT THEOREM
7.1   LIMIT THEOREM FOR INDEPENDENT RANDOM VARIABLES

DEFINITION 1.33 *A sequence $X_n, n \in \mathcal{N}$ of random variables converges in law towards the random variable X if F_{X_n} converges towards F_X pointwise at continuity points of F.*

THEOREM 1.12 *Let $X_1 \ldots X_n$ be a sequence of independent identically distributed random variables whose mean and variance exist, then the sequence: $\frac{S_n - E(S_n)}{\sigma(S_n)}$ converges in law towards the gaussian random variable $N(0,1)$.*

7.2   LIMIT THEOREM FOR DEPENDENT RANDOM VARIABLES

Let X_n, $n \in \mathcal{N}$ be a sequence of random variables. Consider the sequence of σ-fields $\tau_n = \sigma(X_n; m \leq n)$.

DEFINITION 1.34 *The sequence of random variables $X_n, n \in \mathcal{N}$ is a sequence of martingale increments if*

$$E(X_n|\tau_{n-1}) = 0, \ \forall n \in \mathcal{N}.$$

REMARK 4 *Therefore: $E(X_n|\tau_{n-k}) = 0$, $\forall n \in \mathcal{N}$, $1 \leq k \leq n$.*

DEFINITION 1.35 *Let $S_n = \sum_1^n X_n$ and $B_n = E(S_n^2)$. The sequence $X_n, n \in \mathcal{N}$ satisfyes the Lindeberg condition if*

$$\forall \epsilon > 0, \ \frac{1}{B_n} \int_{|X_n| > \epsilon \sqrt{B_n}} X_k^2 dP \to_{n \to \infty} 0$$

where P is the distribution of the random variable X_k.

For simplicity we assume that the random variables X_n, $n \in \mathcal{N}$ are identically distributed.

THEOREM 1.13 *If $X_n, n \in \mathcal{N}$ is a sequence of martingale increments satisfying the Lindeberg condition and if $\frac{1}{B_n} \sum_{k=1}^n X_k^2 \to l$ in probability,* *then*

$$\frac{S_n}{\sqrt{B_n}} \xrightarrow{law} N(0,l), \ as \ n \to \infty.$$

REMARK 5 *Under different asssumptions, the limit distribution is not necessarily gaussian (see [Castillo, 1988], for example).*

8.   LARGE DEVIATIONS

If $X_n, n \in \mathcal{N}$ is a sequence of centered, i.i.d. random variables, μ the common probability distribution (i.e. probability measure), then the law of large numbers tells us that:

$$\forall \epsilon > 0, \ P(|\frac{S_n}{n}| > \epsilon) \to_{n \to \infty} 0$$

and we want now to know the rate of convergence.
The inequalities stated previously give some partial information on upper bounds but no lower bound (out of zero) is provided. The **Large Deviation Principle** fulfils this gap.

THEOREM 1.14 *Assume that the probability measure μ has a Laplace transform $\Psi(t) = E(e^{tX_1})$ and let $\Psi(t) = \ln(\mu\ (t))$ then*

$$P(\frac{S_n}{n} \geq \epsilon) \sim_{n \to \infty} \frac{e^{-n\lambda(\epsilon)}}{a(\epsilon)\sqrt{2\pi n}}$$

where $\lambda(\epsilon) = sup_{t \in R}[t\epsilon - \ln(\mu\ (t))]$, $a(\epsilon) = s\sqrt{\Psi''(s)}$ and s is a zero of $\Psi'(s) = \epsilon$.

REMARK 6 *For explicit computations, it is helpful to mention that*

$$sup_{t \in R}(t\epsilon - \Psi(t)) = s\Psi'(s) - \Psi(s).$$

EXAMPLE 10 *If μ is the gaussian distribution (i.e. measure), we get*

$$P(\frac{S_n}{n} \geq \epsilon) \sim_{n \to \infty} \frac{1}{\sigma\sqrt{2\pi n}} e^{\frac{-n\epsilon^2}{2\sigma^2}}.$$

The Large Deviation Principle has a more refined statement.

THEOREM 1.15 *Under the same assumption as in the previous theorem:*

- $\forall F$, *closed subset of $\mathcal{R}$*

$$limsup_{n \to \infty} \frac{1}{n} ln(P(S_n \in F)) \leq sup_{x \in F}(-\lambda(x))$$

- $\forall F$, *open subset of $\mathcal{R}$*

$$liminf_{n \to \infty} \frac{1}{n} ln(P(S_n \in F)) \geq sup_{x \in G}(-\lambda(x)).$$

9.   STOCHASTIC PROCESSES

We now consider random variables where the time parameter is allowed to vary continously.

DEFINITION 1.36 *A set* $\{X(t,w), w \in \Omega\}$ *of random variables, where* t *is the time parameter is called stochastic process when the parameter set* T *is ordered.*
We usually write simply $\{X(t)\}$.

9.1   BROWNIAN MOTION

Brownian motion is defined by the condition that $X(0) = 0$, and that for $t > s$ the variable $X(t) - X(s)$ is independent of $X(s)$ with a variance depending only on $t - s$.
In other words, the process has independent increments and stationary transition probabilities.

$$E(X^2(t)) = \sigma^2 t$$

and

$$E(X(s)X(t)) = \sigma^2 s \; for \; s < t.$$

For $\tau > t$ the transition densities from (t, x) to (τ, y) are gaussian (i.e. normal) with expectation x and variance $\sigma^2(\tau - t)$, where σ^2 is a constant.

9.1.1   BROWNIAN MOTION WITH ABSORBING (RESP. REFLECTING) BARRIER AT THE ORIGIN

Let q_t be the transition densities of our process. The boundary condition is $q_t(0, y) = 0$, $\forall t$ and $q_t(x, y)$ is given by:

- If the origin is an absorbing barrier:

$$q_t(x, y) = \frac{1}{\sqrt{2\pi t}} [e^{(-\frac{(y-x)^2}{2t})} - e^{(-\frac{(y+x)^2}{2t})}]$$

where $t > 0$, $x > 0$, $y > 0$.

- If the origin is a reflecting barrier:

$$q_t(x, y) = \frac{1}{\sqrt{2\pi t}} [e^{(-\frac{(y-x)^2}{2t})} + e^{(-\frac{(y+x)^2}{2t})}]$$

9.1.2 BROWNIAN MOTION WITH TWO ABSORBING (RESP. REFLECTING) BARRIERS

- Consider a Brownian motion impeded by two absorbing barriers at 0 and $a > 0$.

$$q_t(x, y) = \frac{1}{\sqrt{2\pi t}} \sum_{k=-\infty}^{+\infty} [e^{(-\frac{(y-x+2ka)^2}{2t})} - e^{(-\frac{(y+x+2ka)^2}{2t})}]$$

- if 0 and $a > 0$ are reflecting barriers

$$q_t(x, y) = \frac{1}{\sqrt{2\pi t}} \sum_{k=-\infty}^{+\infty} [e^{(-\frac{(y-x+2ka)^2}{2t})} + e^{(-\frac{(y+x+2ka)^2}{2t})}]$$

9.2 ORNSTEIN-UHLENBECK PROCESS

DEFINITION 1.37 *This is the stochastic process such that for $\tau > t$ the transition density from (t, x) to (τ, y) is gaussian with expectation $e^{-\lambda(r-t)}x$ and variance $\sigma^2(1 - e^{-2\lambda(r-t)})$.*

This process can also be viewed from a particles point of view. It is obtained by subjecting the particles of a Brownian motion to an elastic force (a drift towards the origin of a magnitude proportional to the distance).

References

In this chapter we have presented, without proofs, some elementary probability theory material. For details and complementary material on the subject we warmly recommend [Feller, 1970]. [Hida, 1980] is a more theoretical presentation of Brownian motion, and [Revuz, 1975] presents the full theory of Markov Chains. Large deviations and applications are developed in [Bucklew, 1990, Dembo and Zeitouni, 1993]. The theory of large deviations is explained in [Stroock, 1984] and [Varadhan, 1984].

Chapter 2

STATISTICAL TOOLS

1.   TESTS

1.1   A TEST BASED ON THE CENTRAL LIMIT THEOREM

We explain this test through the following example.

EXAMPLE 11 *An urn contains white and red balls in unknown proportions. We draw n balls and replace them in the urn. The examination of n balls shows a ratio P_0 of white balls. We want to estimate the true ratio p of white balls and express the degree of confidence of the estimate.*

A solution

With each white (resp. red) ball we associate the value 1 (resp. 0). Let $(X_n), n \in \mathcal{N}$ be a sequence of i.i.d. Bernouilli random variables with distribution given by: $P(X_i = 1) = p$, $P(X_i = 0) = 1 - p$ let $S_n = \sum_1^n X_i$. Then we have: $E(\frac{S_n}{n}) = p$ and $var(\frac{S_n}{n}) = \frac{p(1-p)}{n}$.

The central limit theorem tells us that:

$$\forall (a,b) \in \mathcal{R}^2, lim_{n \to \infty} P(a \leq \frac{\frac{S_n}{n} - p}{\sqrt{\frac{p(1-p)}{n}}} \leq b) = \int_a^b \frac{1}{\sqrt{2\pi}} \exp^{-\frac{x^2}{2}} dx$$

Let $\Phi(x) = \int_{-x}^x \frac{1}{\sqrt{2\pi}} e^{-\frac{t^2}{2}} dt$. Then:

$$P(-x \leq \frac{\frac{S_n}{n} - p}{\sqrt{\frac{p(1-p)}{n}}} \leq x) \to_{n \to \infty} \Phi(x).$$

29

With probability close to $\Phi(x)$,

$$\frac{|\frac{S_n}{n} - p|}{\sqrt{\frac{p(1-p)}{n}}} \leq x.$$

But p_0 is a value taken by $\frac{S_n}{n}$. Therefore:

$$|p_0 - p| \leq x\sqrt{\frac{p(1-p)}{n}}$$

In other terms: $(p_0-p)^2 \leq x^2\frac{p(1-p)}{n}$ and p is the solution to the following second order polynomial inequality:

$$(n+x^2)p^2 - (2p_0n + x^2)p + np_0^2 \leq 0$$

whose discriminant is $\Delta = x^4 + 4nx^2p_0(1-p_0)$ and we see that $\Delta \geq 0$, $\forall x \in R$. The two solutions are:

$$a_1 = \frac{2p_0n + x^2}{2(n+x^2)} - x\frac{\sqrt{x^2 + 4np_0(1-p_0)}}{2(n+x^2)},$$

$$a_2 = \frac{2p_0n + x^2}{2(n+x^2)} + x\frac{\sqrt{x^2 + 4np_0(1-p_0)}}{2(n+x^2)}$$

$[a_1, a_2]$ is called the confidence interval of p at level of confidence $\Phi(x)$. When n is large, we can neglect x^2 and we then gets:

$$p_0 - x\sqrt{\frac{p_0(1-p_0)}{n}} < p < p_0 + x\sqrt{\frac{p_0(1-p_0)}{n}}$$

Many applications use $\Phi(x) = 0.95$ leading to $x \approx 1.96$.

1.2   CHISQUARE TEST

DEFINITION 2.1 *Let $X_1,\ldots,X_n$ be a sequence of independent identically distributed random variables with law $N(0,1)$. The chisquare law (χ_n^2) is the law of the random variable $Y_n = \sum_{i=1}^{n} X_i^2$.*

The following proposition summarizes the properties of Y_n.

PROPOSITION 4 $E(Y_n) = n$, $var(Y_n) = 2n$ *and the density f_n of Y_n is given by: $f_n(x) = 0$ if $x > 0$*

$$f_n(x) = \frac{x^{\frac{n}{2}-1}e^{-\frac{x}{2}}}{2^{\frac{n}{2}}\Gamma(\frac{n}{2})} \text{ where } \Gamma(x) = \int_0^{\infty} t^{x-1}e^{-t}dt \text{ for } x > 0$$

REMARK 7 *Recall that:*

$$\Gamma(x+1) = x\Gamma(x) \ for \ x > 0,$$

$$\Gamma(n+1) = n! \ for \ n \in \mathcal{N},$$

and

$$\Gamma(\frac{1}{2}) = \sqrt{\pi}.$$

EXAMPLE 12 *Check if a die is fair.*
A die is fair if all sides show up with probability $\frac{1}{6}$.
Rolling a die 6000 times produces the following results:

Face	1	2	3	4	5	6
Number of times n_i	1034	958	1066	1084	972	886

Check if at the 0.01 level of significance the die is fair.
Solution.
Compute

$$d = \sum_{i=1}^{6} \frac{(n_i - n_i')^2}{n_i'}$$

where $n_i' = 1000$, $i = \{1,\dots,6\}$ and the $n_i's$ are listed above.
We get $d \approx 0.99$. d is the realization of a χ_n^2 random variable and $n = 6 - 1 = 5$. The theoretical value corresponding to χ_5^2 at the 0.01 level of significance is $D \approx 15.086$ (cf. Fig 2.1).
Since $d < D$, d lies within the acceptance region and we conclude that the die is fair at the 0.01 level of significance.

EXAMPLE 13 *A telephone company has recorded (see table below) over 1000 days the number of calls n_i received by a V.I.P. and the corresponding number of days d_i:*

n_i	0	1	2	3	4	5	6	7	8	9	10	11
d_i	14	70	155	185	205	150	115	65	30	5	1	5

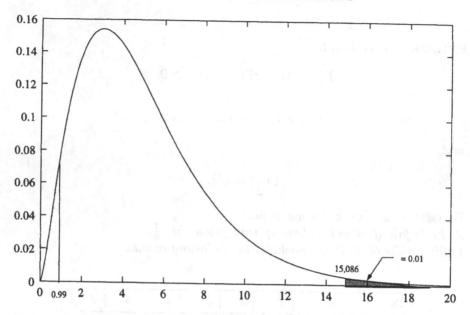

Figure 2.1. The density of χ^2_5. The dashed area represents a probability equal to 0.01.

Check if, at the 0.05 level of significance, the random variable "Number of calls received by the V.I.P." has Poisson distribution.

Left as an exercise.

2.   DISTANCE BETWEEN TWO DISTRIBUTIONS

Given two distributions – for example two states in an HMM – how can we measure the distance between them? In this section we introduce the concept of divergence [Tou and Gonzales, 1974], often known as the Kullback-Leibler measure. In the framework of statistical pattern recognition, the distribution $p_i()$ may represent the dispersion of samples drawn from class ω_i. Let be x, a given sample of unknown class membership. The discriminating information

$$u_{ij} = \ln \frac{p_i(x)}{p_j(x)}$$

is a measure of the information given by the two distributions about the membership of x.

The Kullback-Leibler information number is defined as the expected discriminating information

$$I(i,j) = \int_x p_i(x) \ln \frac{p_i(x)}{p_j(x)} dx.$$

In order to have a symmetric quantity in i and j, we define the divergence $J(ij)$ as:

$$Jij = I(i,j) + I(j,i).$$

When p_i and p_j are gaussian distributions:

$$p_i(x) = \mathcal{N}(\mathbf{x}; \mu_i, \Sigma_i) = \frac{1}{\sqrt{(2\pi)^p \det \Sigma_i}} \exp^{-\frac{1}{2}(\mathbf{x}-\mu_i)^t \Sigma_i^{-1}(\mathbf{x}-\mu_i)}$$

$$p_j(x) = \mathcal{N}(\mathbf{x}; \mu_j, \Sigma_j) = \frac{1}{\sqrt{(2\pi)^p \det \Sigma_j}} \exp^{-\frac{1}{2}(\mathbf{x}-\mu_j)^t \Sigma_j^{-1}(\mathbf{x}-\mu_j)}$$

$$(2.1)$$

the divergence has an analytic form:

$$J_{ij} = \frac{1}{2} \text{tr}[(\Sigma_i - \Sigma_j)(\Sigma_j^{-1} - \Sigma_i^{-1})]$$

$$+ \frac{1}{2} \text{tr}[(\Sigma_i^{-1} + \Sigma_j^{-1})(\mu_i - \mu_j)(\mu_i - \mu_j)^t]$$

In this equation, $\text{tr}(\Sigma)$ denotes the trace of matrix Σ and μ^t the transpose of vector μ.

A special case arises when the two covariance matrices are the same:

$$\Sigma_i = \Sigma_j = \Sigma.$$

In this case J_{ij} becomes the Mahalanobis distance:

$$J_{ij} = (\mu_i - \mu_j)^t \Sigma^{-1} (\mu_i - \mu_j).$$

3.   THE EM ALGORITHM FOR HMM

The EM algorithm was first presented by Dempster in 1977 [Dempster et al., 1977]. This algorithm is an iterative procedure that builds a sequence of models whose parameters converge towards the maximum likelihood estimator. The EM algorithm is valid not only in the field of HMM, but rather in the more general framework of incomplete data. Incomplete data means that the sources of the data are hidden, like the paths of an HMM that produce the sequence of observations. In the following we will restrict ourselves to the HMM framework.

Let θ denote the set of parameters that defines a model. We call Y_t the random variable whose realizations are the observations y_t made at time t. $Y = y$ means $Y_1 = y_1, Y_2 = y_2, \ldots Y_T = y_T$ when we have T observations.

Given an observation sequence y, we are concerned with the maximization of the model's likelihood: $P(Y = y/model)$. We will define $f(y/\theta) = P(Y = y/\theta)$ to express the fact that the likelihood is a function of θ when the observation sequence y is given.

In an HMM, the likelihood is the sum of the probabilities of each time warping alignment between the states $s_1, \ldots s_T$ and the observation sequence $y_1, \ldots y_T$.

Equation

$$f(y/\theta) = \sum_{s \in S^T} g(Y = y, S = s/\theta)$$

can be more conveniently written as:

$$f(y/\theta) = \sum_{s \in S^T} g(y, s/\theta).$$

The log likelihood is defined as:

$$L(y/\theta) = \log f(y/\theta).$$

In a first-order HMM, the probability of a time warping alignment is defined as:

$$g(y, s/\theta) = \prod_t a_{s_t s_{t+1}} b_{s_{t+1}}(y_{t+1}).$$

The maximization of the log likelihood is not easy with the classical downhill directed gradient methods. The EM algorithm is an interesting alternative.

The EM algorithm builds a sequence of models $\theta_0, \theta_1, \ldots \theta_m \ldots$ such that:

$$L(y/\theta_0) \le L(y/\theta_1) \ldots \le L(y/\theta_m)$$

that converges to a stationary point. Under certain conditions of regularity [Dempster et al., 1977], we get a local maximum θ_{MLE}.

The state sequence (= the path) $s_1, s_2, \ldots s_T$ is unknown – there are several with different probabilities — the algorithm replaces $f(y/\theta)$ with the conditional expectation of $g(y, s/\theta)$ computed on a given distribution of paths.

Figure 2.2 shows two distributions of time warping alignments in the context of speech recognition . In each sub-figure, the probabilities $Prob(S_t = s_t, S_{t+1} = s_{t+1}/\theta, y)$ are represented in gray scale. In

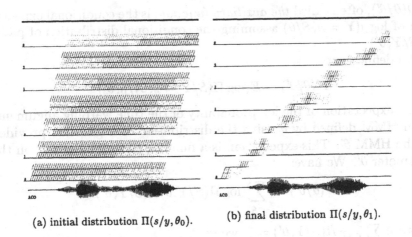

(a) initial distribution $\Pi(s/y, \theta_0)$.   (b) final distribution $\Pi(s/y, \theta_1)$.

Figure 2.2. Two distributions of paths between phonemes and observations in a sentence.

Sub-figure 2.2(a), we have used a IIMM – represented by the parameter θ_0 – where the transitions have the same probabilities and the distributions of observations are uniform; therefore, all paths have the same probability. In the sub-figure 2.2(b), the parameter θ_1 – obtained by the EM algorithm – gives higher probabilities to the path close to the diagonal depending on the speaking rate. Image 2.2(b) can be viewed as an enhanced picture of Sub-figure 2.2(a) which looks less sharp. The EM algorithm can be applied to image restoration as well.

In the following, we will define

$$\Pi(S/Y, \theta) = \frac{g(Y, S/\theta)}{f(Y/\theta)} \qquad (2.2)$$

as the distribution of the hidden paths assuming the model θ and the sequence of observations y.

When there is only one alignment, this distribution is represented by a unique path between the lower left and the upper right corners. This unique path has, obviously, a probability of 1.

The purpose of the EM algorithm is to build iteratively a sequence of distributions $\Pi_0, \Pi_1, \ldots$ in which the dynamic of the time warping alignment probabilities is increasing.

We define, next, the function Q of θ, depending on the parameter θ':

$$Q(\theta/\theta') = E\left(\log g(Y = y, S/\theta)/y, \theta'\right).$$

$Q(\theta/\theta')$, often called *the auxiliary function*, is the conditional expectation of $\log g(Y = y, S/\theta)$ assuming the conditional distribution of paths $\Pi(S/Y, \theta')$.

We define next

$$H(\theta/\theta') = E(\log \Pi(S/Y = y, \theta)/y, \theta')$$

as the expectation of the log probability of the time warping alignments in an HMM defined by θ using the distribution of alignments provided by the HMM θ'. This expectation is a function of θ and depends on the parameter θ'. We have:

$$H(\theta/\theta') = \sum_{s \in S^T} \log \Pi(S/Y, \theta)\Pi(S/Y, \theta').$$

Because $\sum_{s \in S^T} \Pi(S/Y, \theta') = 1$, we get

$$L(y/\theta) = \sum_{s \in S^T} \log f(Y/\theta)\Pi(S/Y, \theta') = E(\log f(Y/\theta)/y, \theta').$$

Using the definition of Π given in equation 2.2, we replace $f(Y/\theta)$ in the above formula and get:

$$L(y/\theta) = E\left(\log\left(\frac{g(Y, S/\theta)}{\Pi(S/Y = y, \theta)}\right)/y, \theta'\right)$$

and then:

$$L(y/\theta) = Q(\theta/\theta') - H(\theta/\theta'). \tag{2.3}$$

The first step of the algorithm – the **E-step**, for Expectation – computes the value of $Q(\theta/\theta')$. The second step – the **M-step**, like Maximization – determines the value of θ for which $Q(\theta/\theta')$ is maximum. Let's call this value θ_m. We have:

$$Q(\theta_m/\theta' = \max_\theta Q(\theta/\theta').$$

Next, we have to use the Jensen inequality that states:

$$H(\theta'/\theta') \geq H(\theta/\theta') \quad \forall \theta.$$

This inequality gives an upper bound for $H(\theta_m/\theta')$. Using this inequality and the definition of θ_m, we can derive from the equation 2.3

$$L(y/\theta_m) \geq L(y/\theta').$$

This property of monotony shows that the algorithm converges to a local optimum that approximates the maximum likelihood estimator [Redner and Walker, 1984]. This convergence is achieved by means of the maximization of $Q(\theta/\theta')$ which is easier to compute than $L(y/\theta)$.

3.1 CONCLUSION FOR THE EM ALGORITHM

Given an initial model, θ_0, the EM algorithm builds a sequence of models $\theta_1, \ldots \theta_{m-1}$ as follows:

step E : we compute $Q(\theta/\theta_{m-1})$

step M : we compute the value θ_m that maximizes $Q(\theta/\theta_{m-1})$.

3.2 APPLICATION TO FIRST ORDER HMM

We have

$$g(y,s) = \prod_t a_{s_t s_{t+1}} b_{s_{t+1}} y(t+1).$$

We call $1_{\{s_t=u, s_{t+1}=v\}}$ the characteristic function of a path s. This function is 0 except if $s_t = u$ and $s_{t+1} = v$ where it is 1.

We can factor the $a_{s_t s_{t+1}}$ and express $g(y,s)$ by means of the characteristic functions

$$g(y,s) = \prod_{(u,v)\in S^2} a_{uv}^{\sum_t 1_{\{s_t=u, s_{t+1}=v\}}} \prod_t b_{s_{t+1}} y(t+1).$$

Let's look more closely at the first term of the product and analyze its contribution in $Q(\theta/\theta_m)$.

$$Q(\theta/\theta_m) = E\left(\log g(Y, S/\theta)/y, \theta_m\right)$$

$$= \sum_{(u,v)\in S^2} E\left(\sum_t 1_{\{s_t=u, s_{t+1}=v\}}/y, \theta_m\right) \log a_{uv} + \cdots \qquad (2.4)$$

The conditional mean of the characteristic function relative to θ_m and y is equal to

$$E(1_{\{s_t=u, s_{t+1}=v\}}/y, \theta_m) = Prob(s_t = u, s_{t+1} = v/y, \theta_m).$$

Then

$$Q(\theta/\theta_m) = \sum_{(u,v)\in S^2} \sum_t Prob(s_t = u, s_{t+1} = v/y, \theta_m) \log a_{uv} + \cdots$$

We maximize $Q(\theta/\theta_m)$ by maximizing each term. Let us examine the first term assuming the constraint:

$$\sum_{v\in S} a_{uv} = 1, \quad \forall u \in S.$$

We introduce the zero quantity — $\lambda(1 - \sum_v a_{uv})$ — called the Lagrange multiplier.

$$Q(\theta/\theta_m) = \sum_{u \in S} \sum_{v \in S} \sum_t Prob(s_t = u, s_{t+1} = v/y, \theta_m) \times \log a_{uv}$$
$$+ \sum_{u \in S} \lambda(1 - \sum_v a_{uv}) + \cdots \tag{2.5}$$

Differentiation with respect to a_{uv} gives

$$\frac{\partial Q(\theta/\theta_m)}{\partial a_{uv}} = \frac{\sum_t Prob(s_t = u, s_{t+1} = v/y, \theta_m)}{a_{uv}} - \lambda = 0.$$

We get the value of a_{uv} as a function of λ and we sum over v, assuming that $\sum_v a_{uv} = 1$. Then, we get the value of λ.

$$\lambda = \frac{1}{\sum_{v \in S} \sum_t Prob(s_t = u, s_{t+1} = v/y, \theta_m)}$$

The value of a_{uv} that maximizes $Q(\theta/\theta_m)$ is:

$$a_{uv} = \frac{\sum_t Prob(s_t = u, s_{t+1} = v/y, \theta_m)}{\sum_v \sum_t Prob(s_t = u, s_{t+1} = v/y, \theta_m)}. \tag{2.6}$$

We maximize the other terms of $Q(\theta/\theta_m)$ in the same way.

The numerator of equation 2.6 can be interpreted as the number of transitions that go from state u to state v during the entire process that produces the sequence of observations y. The purpose of the denominator is to normalize the counts into transition probabilities .

We have to compute $Prob(s_t = u, s_{t+1} = v/y, \theta_m)$. This quantity can be computed recursively on u, v and t.

We define $\alpha(t, u)$:

$$\alpha(t, u) = Prob(s_t = u, y_1, \ldots y_t)$$

as the probability of a partial time warping alignment ending at time t in state u. $\alpha(t, u)$ can be computed iteratively [Baker, 1974, Rabiner and Juang, 1995]:

$$\alpha(t, v) = \sum_u \alpha(t - 1, u) a_{uv} b_v(y_{t+1}). \tag{2.7}$$

We can compute $f(y/\theta) = Prob(y_1, \ldots y_T)$ using $\alpha(t, v)$:

$$Prob(y_1, \ldots y_T) = \sum_{v \in S} \alpha(T, v). \tag{2.8}$$

We, also, define $\beta(t, u)$ as:

$$\beta(t, u) = Prob(t_{t+1}, \ldots y_T / s_t = u)$$

which can be computed recursively as follows:

$$\beta(t, u) = \sum_{v \in S} a_{uv} b_v(y_{t+1}) \beta(t+1, v). \qquad (2.9)$$

By means of the Bayes's formula , it is easy to show [Baker, 1974, Kriouile, 1990]

$$\begin{aligned} Prob(s_t = u, s_{t+1} = v/y) &= \frac{Prob(s_t = u, s_{t+1} = v, y)}{Prob(y_1, \ldots y_T)} \\ &= \frac{\alpha(t, u) a_{uv} b_v(y_t) \beta(t+1, v)}{\sum_{v \in S} \alpha(T, v)}. \end{aligned} \qquad (2.10)$$

Therefore, using Equation 2.6, we have the new estimate of $a(u, v)$ that occurs in the definition of θ_m. The new estimate of $b_v(y(t)$ depends on the *pdf* family: parametric or non parametric. There is an interesting discussion on this topic in Liporace [Liporace, 1982].

4. UNSUPERVISED CLASSIFICATION

Classification means extracting relations between the objects or between the objects and their parameters. Using a measure of proximity or dissimilarity, we aim for the construction of a partition of the set of objects that gives the highest homogeneity to each class with respect to the measure. This process is called *clustering*.

The purpose of clustering is to be efficient and to infer correctly. This goodness depends on the objective of the analysis. But in some cases, the clustering algorithms do not extract classes that reflect the usual concepts assigned to objects. This reason is due to the choice of the distance measure between the objects.

In the following, we will call the objects *the individuals* or *the samples*. The measurements that are carried out on the samples are called *the variables*.

Two kinds of measures can be made on the data. Quantitative variables give information like size, or weight ... The space of these variables is usually $\mathbb{R}^p$, whereas qualitative variables take their values from a finite set, like a set of colors or logical values. In the following, we restrict our attention to the quantitative case, which is the most common case in pattern recognition.

4.1   NOTATIONS

We represent the values taken by the p variables on n samples in an array X of n rows and p columns. Each row is a vector of $\mathbb{R}^p$. This space is called the space of individuals (or samples).

$$X = \begin{pmatrix} x_1^1 & x_1^2 & \cdots & x_1^p \\ x_2^1 & x_2^2 & \cdots & x_2^p \\ \vdots & \vdots & \ddots & \vdots \\ x_n^1 & x_n^2 & \cdots & x_n^p \end{pmatrix} = \left(\{ x_i^j \} \right) , \ i = 1, \cdots n \text{ and } j = 1, \cdots p.$$

To each individual w_i is assigned the i th row of the array defined by the vector:

$$\underline{x}_i = \begin{pmatrix} x_i^1 \\ x_i^2 \\ x_i^3 \\ \vdots \\ x_i^p \end{pmatrix}$$

The set of rows defines the space of individuals. In some cases, an individual or a sample w_i may receive a normalized weight $p_i > 0$ such that $\sum_{i=1}^n p_i = 1$.

Similarly, we introduce $\mathbb{R}^n$ the space of variables: To each variable, v_j is assigned the j th column of the array defined by the vector:

$$\underline{x}^j = \begin{pmatrix} x_1^j \\ x_2^j \\ x_3^j \\ \vdots \\ x_n^j \end{pmatrix}$$

4.2   DEFINITIONS IN THE SPACE OF VARIABLES

Mean:. $\overline{x}^j = \sum_{i=1}^n p_i x_i^j$

Variance:. $\mathrm{var}(\underline{x}^j) = \sum_{i=1}^n p_i (x_i^j - \overline{x}^j)^2 = \sum_{i=1}^n p_i (x_i^j)^2 - (\overline{x}^j)^2$

Standard deviation:. $s(\underline{x}^j) = \sqrt{\mathrm{var}(\underline{x}^j)}$

Covariance:. To the pair of variables corresponding to the vector $\underline{x}^j$ and $\underline{x}^k$, we associate the covariance:

$$\mathrm{cov}(\underline{x}^j, \underline{x}^k). = \sum_{i=1}^{n} p_i(x_i^j - \overline{x}^j)(x_i^k - \overline{x}^k) = \sum_{i=1}^{n} p_i x_i^j x_i^k - \overline{x}^j \overline{x}^k$$

The set of covariances of the $p \times p$ pairs of variables constitutes the covariance matrix of the array. This matrix – p rows and p columns – is symmetric.

$$V = \left(\{\mathrm{cov}(\underline{x}^j, \underline{x}^k)\} \right), \; j = 1, \cdots p \text{ and } k = 1, \cdots p.$$

The p elements of the diagonal are the variances of the p variables.

There is a geometric interpretation of the covariance between two variables of X that can be defined in terms of the scalar product of $\mathbb{R}^n$. Let us consider two vectors on $\mathbb{R}^n$:

$$\underline{x} = \begin{pmatrix} x_1 \\ x_2 \\ x_3 \\ \vdots \\ x_n \end{pmatrix} \text{ and } \underline{y} = \begin{pmatrix} y_1 \\ y_2 \\ y_3 \\ \vdots \\ y_n \end{pmatrix}$$

We can define a scalar product either as follows:

$$< \underline{x}, \underline{y} > = \sum_{i=1}^{n} p_i x_i y_i$$

or in terms of product of matrices:

$$< \underline{x}, \underline{y} > = \begin{pmatrix} x_1 & x_2 & \cdots & x_n \end{pmatrix} \begin{pmatrix} p_1 & 0 & \cdots & 0 \\ 0 & p_2 & \cdots & 0 \\ \vdots & \vdots & \ddots & \vdots \\ 0 & 0 & \cdots & p_n \end{pmatrix} \begin{pmatrix} y_1 \\ y_2 \\ y_3 \\ \vdots \\ y_n \end{pmatrix} = \underline{x}^t D_p \underline{y}$$

Now assume that each column of X is a sample drawn from a centered variable (*i.e.* with zero mean). We get this by the change of variables:

$$\tilde{x}_i^j = x_i^j - \overline{x}^j.$$

If we denote by $\underline{\tilde{x}}^j$ the column j and by $\tilde{X}$ the matrix, then the covariance $\mathrm{cov}(\underline{x}^j, \underline{x}^k)$ can be written:

$$\mathrm{cov}(\underline{x}^j, \underline{x}^k) = < \underline{\tilde{x}}^j, \underline{\tilde{x}}^k >,$$

and the variance becomes:

$$\text{var}(\underline{x}^j) = <\underline{\tilde{x}}^j, \underline{\tilde{x}}^j> = \|\underline{\tilde{x}}^j\|^2.$$

Correlation:.   The correlation is defined as the quotient:

$$cor(\underline{x}^j, \underline{x}^k) = \frac{\text{cov}(\underline{x}^j, \underline{x}^k)}{s(\underline{x}^j)s(\underline{x}^k)}.$$

This is the quotient of the scalar product of the vectors: $\underline{\tilde{x}}^j$ et $\underline{\tilde{x}}^k$ by their norms. It appears to be the *cosine* of the angle between the two vectors. We can put all these quantities in a matrix called the correlation matrix. In this matrix, the elements of the diagonal are 1.

4.3   DEFINITIONS IN THE SPACE OF INDIVIDUALS

In the space $\mathbb{R}^p$, a row represents an individual or a sample. The samples are agglomerated into cells. Let's start by defining the distance between two samples.

4.3.1   DISTANCES BETWEEN SAMPLES

In $\mathbb{R}^p$, one frequently uses distances defined by symmetric, definite and positive matrices (p, p) that generalize the Euclidean distance:

$$d_M(\underline{x}, \underline{y}) = (\underline{x} - \underline{y})^t M(\underline{x} - \underline{y}), \text{ avec } \underline{x} \text{ et } \underline{y} \in \mathbb{R}^p.$$

if M is the identity matrix, $d_M(\underline{x}, \underline{y})$ is the square of the usual Euclidean distance between $\underline{x}$ and $\underline{y}$.

4.3.2   FEATURES OF A CLUSTER

Center of gravity:.   The center of gravity is the centroid of n points considered as elements of $\mathbb{R}^p$ and weighted by p_i:

$$\underline{\overline{x}} = \sum_{i=1}^{n} p_i \underline{x}_i = \begin{pmatrix} \sum_{i=1}^{n} p_i x_i^1 \\ \sum_{i=1}^{n} p_i x_i^2 \\ \vdots \\ \sum_{i=1}^{n} p_i x_i^p \end{pmatrix} = \begin{pmatrix} \overline{x}^1 \\ \overline{x}^2 \\ \vdots \\ \overline{x}^p \end{pmatrix}$$

Inertia:.   The inertia around a point and relative to a particular distance d_M is defined as follows:

$$I_x = \sum_{i=1}^{n} p_i d_M(\underline{x_i}, \underline{x})$$

If $\underline{x}$ is the center of gravity, we define $I_x = I$.

REMARK 8 *The center of gravity is the point that minimizes $I(x) = I_x = \sum_{i=1}^{n} p_i d_M(\underline{x_i}, \underline{x})$.*

Proof: We compute the gradient of $I(x)$ with respect to $\underline{x}$.
If v is a vector of $\mathbb{R}^p$, the gradient relative to v of the quantity:

$$f(v) = v^t M v$$

can be written as:

$$\frac{\partial f(v)}{\partial v} = 2Mv.$$

The minimum is found by equating to zero the partial derivatives:

$$\frac{\partial I(x)}{\partial x} = 2M \sum_{i=1}^{n} p_i(\underline{x} - \underline{x_i}) = \underline{0}$$

then:

$$\sum_{i=1}^{n} p_i(\underline{x} - \underline{x_i}) = \underline{0}.$$

This last equation shows that $\underline{x}$ is the center of gravity of the $\underline{x_i}$.

Cohesion measure of a cell:. Using d_M, we define the cohesion measure of the cell by the quantity:

$$\mu(M) = \sum_{i=1}^{n} \sum_{j=1}^{n} p_i p_j d_M(\underline{x_i}, \underline{x_j}).$$

We can easily show that: $\mu(M) = 2I$.

REMARK 9 *Given a cluster of individuals, among all the matrices M that define a distance and have a determinant of 1, the matrix that minimizes $\mu(M)$ is equal to $(detV)^{\frac{1}{p}} V^{-1}$.*

Proof: The demonstration is given in [Diday et al., 1982] pages 330 – 332.

Variance of a cluster:. The variance of a cluster of samples having different weights is defined as:

$$\text{var}(X) = \frac{1}{\sum_{i=1}^{n} p_i} I.$$

4.3.3   RELATION BETWEEN CENTROIDS:

Let's call E a cluster of samples represented by the matrix X. If there is a partition in k classes $E_1, E_2, \cdots, E_k$, the center of gravity, inertia, and covariance matrices of the sub-cluster E_i share important relationships in classification.

Let $E_1, E_2, \cdots, E_k$ a partition of E. We have:

$$
\begin{array}{ccccc}
 & \overbrace{E_1} & \overbrace{E_2} & & \overbrace{E_k} \\
E = \{ & e_1, e_2, \cdots, e_i, & \cdots, e_l, & , \cdots & \cdots, e_n \} \\
1 = & \underbrace{p_1 + p_2 \cdots p_i+}_{q_1} & \underbrace{\cdots + p_l+}_{q_2} & , \cdots & \underbrace{\cdots + p_n}_{q_k}
\end{array}
$$

Let's define:

$$q_i = \sum_{e_i \in E_i} p_i$$

We define the center of gravity of the cluster E_j by:

$$g_j = \frac{1}{q_j} \sum_i p_i e_i \quad \text{for} \quad e_i \in E_j$$

and the center of gravity of E by:

$$\bar{x} = \sum_{j=1}^{k} q_j g_j.$$

Inter and intra class covariance matrices:. The matrix

$$V_j = \frac{1}{q_j} \sum_{e_i \in E_j} p_i (e_i - g_j)(e_i - g_j)^t$$

is the covariance matrix of E_j. By definition, we call W (like *within*) the intra class matrix defined as the weighted mean of the V_j:

$$W = \sum_{j=1}^{k} q_j V_j.$$

Similarly, we define the inter class matrix B (like *between*) as the covariance matrix of the center of gravity g_j weighted by the mass q_j:

$$B = \sum_{j=1}^{k} q_j(g_j - \overline{x})(g_j - \overline{x})^t.$$

This matrix $(p \times p)$ is not invertible because the k centers of gravity lie in a subspace of dimension $k - 1 < p$. We can show easily:

$$V = W + B.$$

Inter class inertia and intra class inertia:. The Koenig-Huygens theorem gives the relationship between the total inertia and the inertia of each class relatively to the center of gravity:

$$I(E) = \sum_{j=1}^{k} I(E_j) + \sum_{j=1}^{k} q_j d_M(g_j, \overline{x}).$$

4.3.4 DISTANCES BETWEEN CLUSTERS

Inter-group distances:. Figures 2.3(a), 2.3(b) et 2.3(c) illustrate three usual distances between two groups of individuals.

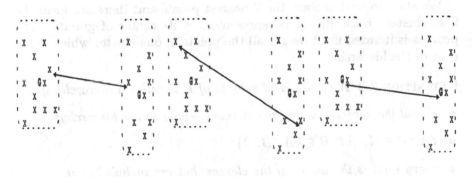

(a) Minimum link   (b) Maximum link   (c) Between centroids

Figure 2.3. Distances between groups

Classes modeled by normal densities:. In this case, the divergence concept between the class is a measure of the distance between the two classes (cf. page 32).

4.4   CLASSIFICATION BY HIERARCHY AND TREE

We are interested in representing the points $w_1, w_2, \ldots, w_n$ by a set of hierarchically nested parts. For example, the hierarchy shown in Figure 2.4(b) represents the relations between the points drawn on Figure 2.4(a), when we use the Euclidean distance.

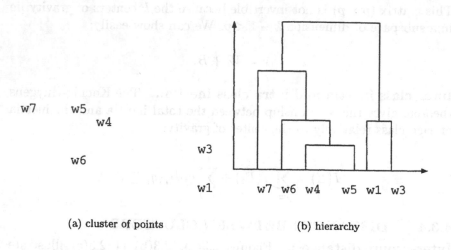

(a) cluster of points   (b) hierarchy

Figure 2.4.   Classification and hierarchy (adapted from [Diday et al., 1982])

We start to agglomerate the 2 nearest points and therefore form the first cluster. Each cluster is represented by its center of gravity. The process is iterated until we get all the points in one cluster which is the root of the hierarchy.

DEFINITION 2.2 *A family H of subsets of E is called a hierarchy if:*

- *E and the subsets having one element belong to the hierarchy;*

- $\forall A, B \in H, A \cap B \in \{A, B, \emptyset\}$*;*

- *every class is the union of the classes that are included in it.*

4.4.1   AGGREGATION CRITERIA

The first step in designing a hierarchy is to choose a similarity measure between the samples, or between the groups of samples and between the groups of samples and the samples. This measure is defined with the help of the distances between samples, groups and between a group and a sample according to the weight of each group. Various criteria can be used:

- Figure 2.3(a) illustrates what is the minimum link between the 2 parts h_1 and h_2:

$$\delta(h_1, h_2) = \min_{w_i \in h_1, \ w_j \in h_2} d(w_i, w_j)$$

- the maximum link is illustrated in Figure 2.3(b):

$$\delta(h_1, h_2) = \max_{w_i \in h_1, \ w_j \in h_2} d(w_i, w_j)$$

- the average of the distances between each pair of samples is illustrated in Figure 2.3(c)

$$\delta(h_1, h_2) = \frac{1}{|h_1||h_2|} \sum_{w_i \in h_1, \ w_j \in h_2} d(w_i, w_j)$$

with $|h_i|$ the cardinal of h_i.

- the distance between the centers of gravity

$$\delta(h_1, h_2) = d(G(h_1), G(h_2)) \tag{2.11}$$

with $G(h_i)$ the center of gravity of h_i

- the variation of the inertia

$$\delta(h_1, h_2) = \frac{p(h_1)p(h_2)}{p(h_1) + p(h_2)}$$

This criteria is frequently used. We have seen, earlier, that the inertia of a cluster is proportional to the cohesion of the cluster. We use this feature to look for the clusters whose aggregation will minimize the variation of the intra-cluster inertia.

- the variation of the variance

$$\delta(h_1, h_2) = \frac{p(h_1)p(h_2)}{(p(h_1) + p(h_2))^2} d(G(h_1), G(h_2))$$

4.4.2 WARD'S ALGORITHM

This algorithm performs a bottom up classification using the criteria of the variation of the variance.

This algorithm is fast. The time involved in the computation of $\delta(h_i, h_j)$ decreases each time two clusters are merged. It is possible to stop when a determined number of cluster or an inertia loss are reached. The main disadvantage of this algorithm lies in its memory requirement because a symmetric matrix $\delta(h_1, h_2)$ as huge as the number of samples must be stored.

Algorithm 2.1 The Ward's algorithm.

1. Start with clusters having one element, each of them having equal weights $p(h_i) = 1$;

2. determine the couple of clusters (h_1, h_2) whose $\delta(h_1, h_2)$ is minimal ;

3. merge the 2 clusters and build the resulting cluster having the sum of weights $p(h_1) + p(h_2)$;

4. keep iterating until all the clusters have been merged.

4.5   DIVISIVE CLASSIFICATION

In this technique, we start by considering all the samples as part of one initial cluster. Then, we proceed by splitting the cluster and looking for the optimal partition into two clusters. At this point, two decisions can be made. Either to split each cluster and classify the set of samples as already before and get 2^l clusters after l iterations [Buzo et al., 1980], or to split only one cluster (for example, the largest, the one having the larger variance, or whatever [Wilpon and Rabiner, 1985].)

Each cluster is represented by a centroid or by a set of samples regularly dispatched in the cluster [Diday et al., 1982].

The optimal classification of the samples into a specified set of L clusters is known as the *K-means* algorithm (described in 2.2). In this algorithm, we call $d(x, g_i)$ the distance between the sample x and the cluster g_i represented by its center of gravity. The demonstration that this algorithm converges can be found in numerous books on statistics and data analysis [Diday et al., 1982, Saporta, 1990]. The demonstration is based upon the fact that the average of the variances computed on the clusters that define a partition decreases at each iteration. The average of the variances is called the *distortion*.

An example is given in Figure 2.6.

4.5.1   EXAMPLE

Figure 2.6 shows the behavior of a divisive program called *LBG* (for Linde Buzo and Gray that described it in [Linde et al., 1980]) on the input data generated by the program given Page 50.

The training samples to be clustered are represented in Fig. 2.5. This figure is made up of 4 cells obtained by simulation using normal population in the x axis and the y axis.

The following fragment of the program generates these samples with the values: NCL = 4, NCOEF = 2, NPTS = 10000. The routine gasdev()

Algorithm 2.2 The K-means algorithm

1. Classify each sample x into the nearest cluster by computing $d(x, g_i)$ $1 < i < L$, choose the cluster where $d(x, g_i)$ is minimal;

2. if a cluster becomes empty, remove it;

3. determine the center of gravity of each new cluster;

4. compute the variance of each cluster and the average of variances over the set of clusters and then compare its average to the average obtained in the former partition. If the decrease is below a given threshold, then stop, otherwise start again in 1.

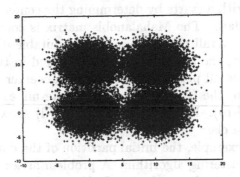

Figure 2.5.   Input data from a 4 class random generator.

comes from the Recepices library. Its generates random samples following a normal distribution with mean and std-dev. equal to 1.

```
static int idnum = -13 ; /* seed for random gener. */
static float Mean[NCL][NCOEF] =
{{10.,   0.},
 { 0.,  10.},
 { 0.,   0.},
 {10.,  10.}
} ; /* mean */
static float Std[NCL][NCOEF] =
{{2.,  2.},
 {2.,  2.},
 {2.,  2.},
 {2.,  2.}
} ; /* standard deviation */
float samples[NCL * NPTS][NCOEF] ; /* input data */
...
```

```
int nSamples ;
for (int ncl = nSamples = 0 ; ncl < NCL ; ncl++)
  for (int npt = 0 ; npt < NPTS ; npt++) {
    for (int i = 0 ; i < NCOEF ; i++)
      samples[nSamples][i] = Mean[ncl][i] +
        Std[ncl][i] * gasdev(&idnum);
    nSamples++ ;
}
```

The *LBG* algorithm starts by determining the center of gravity (centroid) of all the data. The Mahalanobis matrix is computed from the covariance of the overall cluster to specify a suitable distance between the samples. Next, the algorithm splits this cloud in two parts by applying a random small perturbation around the center of gravity (Figure 2.6(a)). Then, the K-means algorithm determines the 2 new centroids (Figure 2.6(b)). Fig. 2.6(c) and Fig. 2.6(d) show the initial and final partition in 4 class.

In this simple example, the initial partition of the data has been discovered by the clustering algorithm. A problem arises when we try to cluster the input data drawn from a 3 class random generator (NCL = 3). Figure 2.7 shows the optimal partition found by the algorithm. It is obvious that a local optimum has been reached. After several iterations, we see that a cluster has been outrageously split. Contrary to the Ward's algorithm, which is deterministic, this top down algorithm gives different results depending on the perturbations that are performed during the splitting process. The final result gives a local minimization of the intra-cluster variance(cf. page 44). To avoid this phenomenon, an intuitive idea is to run successively a top-down clustering algorithm, a bottom up clustering algorithm like Ward's algorithm and, finally, the K-means algorithm to consolidate the overall clusters [Lebart et al., 1995]. This allows to merge the clusters that have been unreasonably split and decreases the over splitting process in the cluster extraction and the over parameterization risk. Wong [Wong, 1982] calls this method the *hybrid method*.

In these figures, we see that all the borders between the clusters are straight lines whose equations are linear in x and y (x and y define the axis). There are other methods of classification, based on automatic neural networks, in which the borders between the clusters are polynomial in x and y.

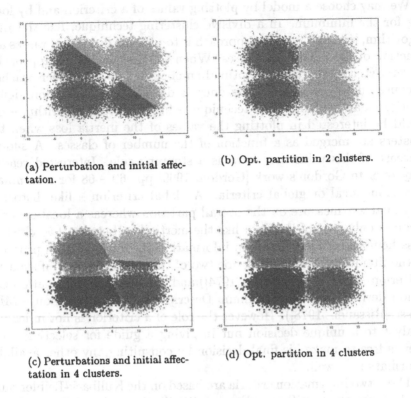

(a) Perturbation and initial affectation.

(b) Opt. partition in 2 clusters.

(c) Perturbations and initial affectation in 4 clusters.

(d) Opt. partition in 4 clusters

Figure 2.6. Top down Classification

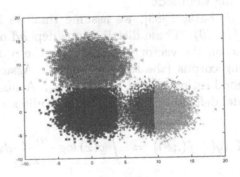

Figure 2.7. Final clustering in the 3 class problem.

4.6 DETERMINING THE UNKNOWN NUMBER OF CLASSES

Determining the unknown number of classes to look for in a clustering analysis is related to the more general problem of model selection. It is still an open area of research in classification and data analysis.

We may choose a model by plotting values of a criterion and by looking for the minimum. In a divisive clustering technique, like the LBG algorithm, the first intuitive approach is to plot the distortion values as a function of the number of classes. When the first derivative approaches to zero, we consider that an optimal number of classes has been reached, because adding one more class doesn't decrease the distortion significantly. In an agglomerative technique – like the Ward algorithm – one could be interested in plotting the values of the inertia loss when two clusters are merged as a function of the number of classes. A sudden increase of this value can serve as a stopping rule. Interested readers may refer to Gordon's work [Gordon, 1999] pp. 60 – 68 for a summary of various local or global criteria. A global criterion – like distortion – evaluates a measure on the overall partition whereas a local criterion involves only part of the data like the inertia loss between two clusters. Less heuristic criteria based on information theory have been proposed in the literature. We will describe two of them: the Akaike Information Criterion – AIC – (see [Akaike, 1974]) and the Rissanen Information Criterion derived from the Minimum Description Length criterion – MDL – (see [Rissanen, 1978]). However the role of a criterion is not in leading analyst to a unique decision but in giving a guide for selection. The user is free to make his final decision by consulting any other available information as well.

These two information criteria are based on the Kullback-Leibler number between the true (but unknown) distribution $g(.)$ and the predictive distribution defining the model.

Like Shibata [Shibata, 1986], we assume that we have a family F of distributions $f(y_n; \theta)$. These distributions depend on the parameter vector $\theta \in \mathbb{R}^p$ and on the vector y_n made with the n observations that define the training corpus (the samples space). Assume that $\hat{\theta}$ is the maximum likelihood estimate of θ based on y_n. A possible measure of how well the model defined by $\hat{\theta}$ fits $g()$ is the Kullback-Leibler measure

$$K_n(g(.), f(.; \hat{\theta})) = \int g(x) \log \frac{g(x)}{f(x; \hat{\theta})} dx.$$

The quantity depends on y_n though $\hat{\theta}$, hence the subscript n. We denote by EK_n the expectation of the Kullback-Leibler measure with respect to the vector of random variables y_n.

Shibata [Shibata, 1986] shows that EK_n can be written as:

$$EK_n = \int g(x) \log(x) dx + E(-l(\hat{\theta})) + A(n, \theta)$$

where $l(\hat{\theta})$ is the log-likelihood estimate of $l(\theta) = \log f(y_n; \theta)$ and $A(n, \theta)$ is a term that controls the amount of penalty for the increasing size of the model – p – and the training corpus size n.

Therefore, in order to minimize EK_n, we have to minimize the last two terms.

When the true distribution $g(.)$ belongs to the family F, $A(n, \theta)$ becomes simpler and is equal to p. This is the Akaike criterion [Akaike, 1974]:

$$AIC = E(-l(\hat{\theta})) + p.$$

Schwarz [Schwarz, 1978] proposes an other criterion called BIC in which a correcting factor $\log n$ increases the penalty when the corpus size increases:

$$BIC = E(-l(\hat{\theta})) + p\log n.$$

This last criterion is similar to the Rissanen [Rissanen, 1978] criterion.

4.6.1   THE AKAIKE AND RISSANEN CRITERIA

In the framework of clustering techniques, we propose to use two criteria as stopping rules: Akaike and Rissanen. We call c the number of clusters. Each sample belongs to $\mathbb{R}^p$. Therefore the number of independant components in θ is at least $c \times p$. The quantity $\frac{E(-l(\hat{\theta}))}{n}$ is estimated by the average distortion given by the K-means algorithm during its last iteration. These criteria may be plotted as a function of the number of classes to help the analyst to make a decision:

Akaike criterion $AIC = distortion + \frac{c \times p}{n}$;

Rissanen criterion $BIC = distortion + \frac{c \times p \times \log n}{n}$.

II
APPLICATIONS

Chapter 3

SOME APPLICATIONS IN ALGORITHMICS

1. PROBABILISTIC ANALYSIS OF QUICKSORT

1.1 QUICKSORT ALGORITHM

Quicksort is a fast sorting algorithm, widely used for internal sorting. The basic idea is the choice of a partitioning element K. For example, let us consider the integer sequence [Régnier, 1989]:

$$45, \ 677, \ 98, \ 43, \ 42, \ 41, \ 60, \ 130, \ 32, \ 67$$

and choose $K = 67$ as the partitioning element. Scanning the left sublist from left to right, one exchanges any key greater than K with a key of the right sublist, scanned from right to left. This builds a list where K has its final position, all the keys to its left (resp. to its right) being smaller (resp. larger) than K. The intermediate stages are:

45	32							677	67
45	32	60				98	130	677	67
45	32	60	43	42	41	98	130	677	67
45	32	60	43	42	41	98	130	677	67

Then the process can be applied recursively to both sublists $(45 \ldots 41)$ and $(130 \ldots 98)$. These successive stages can be represented by a binary search tree. Each key is used as a partitioning element in a recursive call and the level of this call is also the depth of the node containing that key. Now, to evaluate the cost running Quicksort on a data set, we

57

count the number of comparaisons to be performed. Each key in a node is compared once to all the keys occuring in the path from the root to that node. Thus our cost is equal to the External Path Length of the binary search tree built. As the operations of the algorithm only depend on the relative order of the keys, this data set can also be considered as a permutation in σ_n. We make the standard assumption of a uniform distribution of data:

For data sets of size n, the $n!$ relative orders are equally likely. Under this model, the analysis of Quicksort can be reduced and performed on binary search trees built by successive insertions. This standard algorithm for binary search trees is fully described in [Knuth, 1973].

1.2   A STRAIGHTFORWARD AVERAGE CASE ANALYSIS

The expected number M_n of key comparaisons satisfies:

$$M_0 = 0$$

and for $n \geq 1$

$$M_n = \frac{1}{n}[n(n-1) + \sum_{k=1}^{n} M_{k-1} + M_{n-k}]$$

This recurrence formula is easy to solve. Below we give an elementary solution.

M_n writes in the form:

$$nM_n = n(n-1) + 2\sum_{k=1}^{n-1} M_k \qquad (3.1)$$

Replace n by n-1:

$$(n-1)M_{n-1} = (n-1)(n-2) + 2\sum_{k=1}^{n-2} M_k \qquad (3.2)$$

Then substract Eq.(3.1) and Eq.(3.2):

$$nM_n - (n-1)M_{n-1} = 2(n-1) + 2M_{n-1}$$

$$nM_n = (n+1)M_{n-1} + 2(n-1)$$

It follows easily that:

$$M_n = 2(n+1)H_n - 4n$$

where H_n stands for the harmonic sum: $H_n = \sum_{i=1}^{n} \frac{1}{i}$ for all integers $i \geq 1$, and $H_0 = 0$. Since $H_n = \ln(n) + \gamma + O(1/n)$ as $n \to \infty$, where γ is Euler's constant, namely $0.5772156649\ldots$, it follows that as $n \to \infty$,

$$M_n = 2n\ln(n) - (4 - 2\gamma)n + 2\ln(n) + O(1)$$

Now we want to know if there is a strong concentration of probability around the mean (i.e. if events leading to worst case cost are rare events).

Computing the variance of the number of key comparaisons is more intricate with a combinatorial approach [Sedgewick, 1985]. We wonder also if the number of key comparaisons has a limiting distribution. We will see in the next section that the formulation in terms of matingale leads to short easy proofs and shows that Quicksort behaves badly (i.e. $\frac{n^2}{2}$ comparaisons are needed) very seldom.

1.3   A MARTINGALE ASSOCIATED TO QUICKSORT

The formulation in terms of martingale (cf. page 19) has been given by M. Régnier [Régnier, 1989] and U. Rösler [Rösler, 1989]. Our exposition is based on [Régnier, 1989]. We first state some notations.

The depth of insertion of a record is the depth of a node where this record is inserted, where we take the depth of the root to be 0. Let X_n be the random variable counting the depth of insertion of a random record in a binary search tree of size $n - 1$. It's obvious that $X_1 = 0$ and $X_2 = 1$.

The internal path length is: $IPL_n = \sum_{i=1}^{i=n} X_i$. For convenience we define:

$$Y_n = IPL_n + 2$$

The nice result stated below has been proven in [Régnier, 1989].

THEOREM 3.1 *The random variables*

$$Z_n = \frac{Y_n - 2(n+1)(H_{n+1} - 1)}{n+1} = \frac{IPL_n - 2(n+1)H_n + 4n}{n+1}$$

form a martingale with a null expectation. Their variances satisfy:

$$var(Z_n) = 7 - \frac{4\pi^2}{6} + O(1/n)$$

Proof: The proof (given for completeness) is based on the following two lemmas.

LEMMA 1 *Let $E(X_n|X_1,\ldots,X_n)$ and $E(Y_n|X_1,\ldots,X_n)$ be the conditional expectations of X_n and Y_n with respect to $(X_1,\ldots,X_n)$. They satisfy:*

$$E(Y_n|X_1,\ldots,X_n) = \frac{Y_{n-1}}{n}E(Y_n|X_1,\ldots,X_n)$$
$$= \frac{n+1}{n}Y_{n-1} + 2.$$

LEMMA 2 *The expectation of the internal path length satisfy:*

$$E(IPL_n) = 2(n+1)H_n - 4nE(Y_n)$$
$$= 2(n+1)(H_{n+1} - 1).$$

Before going into the proofs of these lemmas, let us show how we derive the results of the theorem. From the two previous lemmas it follows that:

$$E(Z_n|X_1,\ldots,X_{n-1}) = E(\frac{Y_n - 2(n+1)(H_{n+1}-1)}{n+1}|X_1,\ldots,X_{n-1})$$
$$= \frac{Y_{n-1}}{n} + \frac{2}{n+1} - \frac{E(Y_n)}{n+1}$$
$$= Z_{n-1}$$

Therefore $(Z_n), n \in \mathcal{N}$ is a martingale. To derive the expression of the variances, we rewrite:

$$Z_n = \frac{n}{n+1}Z_{n-1} + \frac{X_n - E(X_n)}{n+1}$$

and

$$E(Z_n^2) = \frac{n^2}{(n+1)^2}E(Z_{n-1}^2) + \frac{var(X_n)}{(n+1)^2} + \frac{2n}{n+1}{}^2 E(Z_{n-1}(X_n - E(X_n)))$$

But since $E(X_n - E(X_n)|X_1,\ldots,X_{n-1}) = Z_{n-1}$, we get:

$$\frac{n+1}{n+2}E(Z_n^2) = \frac{n}{n+1}E(Z_{n-1}^2) + \frac{var(X_n)}{(n+1)(n+2)}$$

The moments of the depth of insertion are well known [Knuth, 1973, Louchard, 1987]. In particular:

$$var X_n = 2(H_n - 1) - 4(H_n^{(2)} - 1)$$

with $H_n^{(2)} = \sum_{i=1}^n \frac{1}{i^2} \sim_{n\to\infty} \frac{\pi^2}{6}$.

Summing these expressions yields:

$$E(Z_n^2) = \frac{n+2}{n+1}(7 - 4\frac{\pi^2}{6}) + O(\frac{1}{n})$$

This completes the proof of our theorem.

1.4   A LIMITING DISTRIBUTION

To the martingale (Z_n), $n \in \mathcal{N}$ we can apply Theorem (convergence of matingales). We get that:

$$Z_n \to_{n\to\infty} Z$$

in law and in L^2. We may now characterize the law of Z. This has been done by U. Rösler in [Rösler, 1989]:

THEOREM 3.2 $\frac{X_n - E(X_n)}{n}$ *converges weakly to some random variable* Y. *The distribution of* Y *is characterized as the fixed point of a contraction. It satisfies a recursive equation (which allows numerical approximation). In addition* Y *has exponential tails.*

We skip the proof of this result which is very mathematically involved (see [Rösler, 1989] for details). A consequence of this result is the following estimate of probability:

$$P(X_n \geq 2E(X_n)) \leq C.p.\ln^{-p}n, \ 1 \leq p \leq \infty, \ n \in \mathcal{N}$$

where C is a constant.

REMARK 10 *McDiarmid and Hayward [McDiarmid, 1996] have given a complete analysis of quicksort in terms of large deviations (cf. page 25). They proved the following nice result:*
For fixed ϵ,

$$P(|\frac{X_n}{EX_n - 1}| > \epsilon) = n^{-(2+o(1))\epsilon \ln(\ln(n))}, \ n \to \infty.$$

2.   PROBABILISTIC ANALYSIS OF DYNAMIC ALGORITHMS

We consider data structures subject to the following operations.

I for insertion,

D for deletion,

Q^+ for positive query (i.e. successful search),

Q^- for negative query (i.e. negative search).

The size is unknown and this makes a big difference with the usual situation. Operations like lazy deletions and batched insertions have also been considered. The corresponding dynamic algorithms have been analyzed in [Kenyon-Mathieu and Vitter, 1991, Louchard et al., 1997, Wyk and Vitter, 1986]. How can we analyse such algorithms? Great care is required as shown in the following very simple example provided by Jonassen and Knuth [Jonassen and Knuth, 1978].

2.1   A TRIVIAL ALGORITHM WHOSE ANALYSIS ISN'T

Consisder three ordered elements $x < y < z$, we get five binary search trees on these elements:

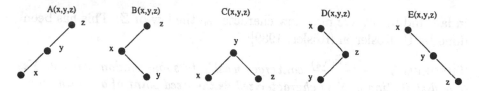

Figure 3.1.   Binary search trees of size 3

(leaves are not drawn)
and the two possibilities on two elements $x < y$ are

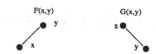

Figure 3.2.   Binary search trees of size 2

The standard insertion algorithm produces the following binary search tree when inserting element z into a tree containing x and y:

Initial tree	Result if $z < x$	Result if $x < z < y$	Result if $y < z$
F(x,y)	A(z,x,y)	B(x,z,y)	C(x,y,z)
G(x,y)	C(z,x,y)	D(x,z,y)	E(x,y,z)

The deletion algorithm operates as follows on a 3-element tree:

Initial tree	Delete x	Delete y	Delete z
A(x,y,z)	F(y,z)	F(x,z)	F(x,y)
B(x,y,z)	F(y,z)	F(x,z)	G(x,y)
C(x,y,z)	G(y,z)	F(x,z)	F(x,y)
D(x,y,z)	G(y,z)	G(x,z)	G(x,y)
E(x,y,z)	G(y,z)	G(x,z)	G(x,y)

If we insert three elements $x < y < z$ in random order, we get a tree of shape A, B, C, D, E with the respective probabilities 1/6, 1/6, 2/6, 1/6, 1/6; then a random deletion leaves us with the following six possibilities and probabilities:

F(x,y)	F(x,z)	F(y,z)	G(x,y)	G(x,z)	G(y,z)
3/18	4/18	2/18	3/18	2/18	4/18

The probability of shape F at this point is $9/18 = 1/2$. Now comes another random insertion, say w. The probability is 1/4 that w is the smallest of $\{w, x, y, z\}$; and the other three cases also occur with probability 1/4. Thus the tree $F(x, y)$ becomes $A(w, x, y)$, $B(x, w, y)$ or $C(x, y, w)$ with respective probabilities 1/4, 1/4, 1/2; and the other cases $F(x, z)$, $\ldots$, $G(y, z)$ can be worked out similarly. One finds that the insertion of w produces a tree of shape A, B, C, D, E with respective probabilities:

$$11/72, \ 13/72, \ 25/72, \ 11/72, \ 12/72$$

A random deletion now produces a tree of shape F with probability:

$$11/72 + (2/3).(13/72) + (2/3).(25/72) > 1/2$$

Uniformity is not preserved. A deep study of this phenomenon has been done by Knott [Knott, 1975], and Knuth [Jonassen and Knuth, 1978, Knuth, 1977]. In fact we realize the following sequences of insertions I and deletions D:

$$III, \ IIIDI, \ IIIDIDI, \ \ldots, \ III(DI)^n$$

So we must study the following trivial procedure:
Step 1

Let x, y be independent uniform random numbers. Insert x into an empty tree, then insert y. (If $x < y$, we get the tree $G(x, y)$, otherwise we get $F(x, y)$.)

Step 2

Insert a new independent uniform random number into the tree.

Step 3

Choose one of the three elements in the tree at random, each with equal probability and delete it.

Step 4

Return to Step 2.

At the beginning of the $(n + 1)$st occurrence of Step 3, we have a tree of shape A, B, C, D, E with probabilities a_n, b_n, c_n, d_n, e_n which we want to compute. The first two times we get step 3, we have seen that $(a_n, \dots, e_n)$ are respectively

$$(1/6, 1/6, 2/6, 1/6, 1/6) \text{ and } (11/72, 13/72, 25/72, 11/72, 12/72)$$

The behavior of this algorithm depends only on the relative order of the elements inserted, and on the particular choice made at each deletion step. Therefore one way to analyze the situation after the pattern $III(DI)^n$ is to consider $(n + 3)!3^n$ configurations to be equally likely, reflecting the relative order of the $n + 3$ elements inserted and the n 3-way choices of which element to delete. However, such a discrete approach leads to great complications. The following continuous approach proposed in [Jonassen and Knuth, 1978] follows the algorithm more closely and turns out to be much simpler.

Let $f_n(x, y)dxdy$ be the (differential) probability that the tree is $F(X, Y)$ at the beginning of Step 2, after n elements have been deleted, where $x \leq X < x + dx$ and $y \leq Y < y + dy$ and let $f_n(x, y)dxdy$ be the corresponding probability that is $G(X, Y)$. It is possible to list recurrence relations for these probabilities a_n, b_n, c_n, d_n, e_n. First we have:

$$a_n(x, y, z) = f_n(y, z),$$

$$b_n(x, y, z) = f_n(x, z),$$

$$c_n(x, y, z) = f_n(x, y) + g_n(y, z),$$

$$d_n(x, y, z) = g_n(x, y),$$

$$e_n(x, y, z) = g_n(x, y)$$

for $0 \leq x < y < z \leq 1$, by considering the six possible actions of Step 2.

REMARK 11 *These probabilities are zero when $x < 0$, $x > y$, $y > z$ or $z > 1$; at the boundaries $x = 0$, $x = y$, $y = z$, and $z = 1$ there may be discontinuities, and it does not matter how we define the functions there.*

Secondly we have

$$
\begin{aligned}
f_{n+1}(x,y) = {} & \frac{1}{3} \int_0^x (a_n(t,x,y) + b_n(t,x,y))dt \\
& + \frac{1}{3} \int_x^y (a_n(x,y,t) + b_n(x,t,y) + c_n(x,y,t))dt \\
& + \frac{1}{3} \int_y^1 (a_n(x,y,t) + c_n(x,y,t))dt,
\end{aligned}
$$

$$
\begin{aligned}
g_{n+1}(x,y) = {} & \frac{1}{3} \int_0^x (c_n(t,x,y) + d_n(t,x,y) + e_n(t,x,y))dt \\
& + \frac{1}{3} \int_x^y (d_n(x,t,y) + b_n(x,t,y) + e_n(x,t,y))dt \\
& + \frac{1}{3} \int_y^1 (b_n(x,y,t) + d_n(x,y,t) + e_n(x,y,t))dt
\end{aligned}
$$

for $0 \leq x < y \leq 1$, by considering the possible actions of Step 3. Inserting the first equations into the second and applying simplifications, we get the fundamental recurrences

$$
\begin{aligned}
f_{n+1}(x,y) = {} & \frac{1}{3}(f_n(x,y) + \int_0^y (f_n(t,y)dt + \int_x^y (f_n(x,t))dt \\
& + \int_x^y g_n(t,y)dt + \int_y^1 f_n(y,t))dt + \int_y^1 g_n(y,t)dt),
\end{aligned}
\tag{3.3}
$$

$$
\begin{aligned}
g_{n+1}(x,y) = {} & \frac{1}{3}(g_n(x,y) + \int_0^x f_n(t,x)dt + \int_0^x g_n(t,y)dt \\
& + \int_0^x g_n(t,x)dt + \int_x^1 g_n(x,t)dt + \int_y^1 f_n(x,t)dt)
\end{aligned}
\tag{3.4}
$$

for $0 \leq x < y \leq 1$.

Consideration of Step 1 also leads to the obvious conditions:

$$
f_0(x,y) = g_0(x,y) = 1
\tag{3.5}
$$

for $0 \leq x < y \leq 1$.

The quantities of interest to us are:

$$a_n = \int_0^1 \int_0^x \int_0^y a_n(X, Y, Z)dxdydz,$$

$\ldots$,

$$e_n = \int_0^1 \int_0^z \int_0^y e_n(X, Y, Z)dxdydz,$$

and

$$f_n = \int_0^1 \int_0^y f_n(x, y)dxdy, \quad g_n = \int_0^1 \int_0^y g_n(x, y)dxdy.$$

In addition $f_0 = f_1$ and $g_0 = g_1$.

When the algorithm reaches Step 2, the two numbers X and Y in its tree are random, except for the condition that $X < Y$. Thus we must have

$$f_n(x, y) + g_n(x, y) = 2 \ for \ 0 \leq x < y \leq 1 \ and \ n \geq 0 \qquad (3.6)$$

It is 2, not 1 since the probability that $x \leq X < x + dx$ and $y \leq Y < y + dy$ given that $X < Y$ is $2dxdy$.

Relation Eq.(3.6) means that we really have only one function to worry about, namely $f_n(x, y)$. We rewrite Eq.(3.3), Eq.(3.4) and Eq.(3.5) to take account of this fact:

$$f_0(x, y) = 1,$$

$$f_{n+1}(x, y) = \frac{1}{3}(2 - 2x + f_n(x, y) + f_n(x, y) + \int_0^x f_n(t, y)dt$$

$$+ \int_x^y f_n(x, t)dt) \ for \ n \geq 0.$$

The first few f_n's are:

$$f_1(x, y) = 1 - \tfrac{2}{3}x + \tfrac{1}{3}y,$$

$$f_1 = \tfrac{1}{2},$$

$$f_2(x, y) = 1 - \tfrac{8}{9}x + \tfrac{4}{9}y + \tfrac{1}{18}(x - y)^2,$$

$$f_2 = \tfrac{109}{216}.$$

f_1 and f_2 correspond to what has been obtained so far. We are hoping that the process converges for large n, and in this case the limiting distribution f_∞ will have to satisfy

$$f_\infty(x,y) = \frac{1}{3}(2 - 2x + f_\infty(x,y) + f_\infty(x,y) + \int_0^x f_\infty(t,y)dt$$
$$+ \int_x^y f_\infty(x,t)dt)$$

The answer is yes but we are skipping this technical proof (see [Jonassen and Knuth, 1978] for details). Let $q(x,y) = f_\infty(2x, 2y)$ then

$$q(x,y) = 1 - 2x + \int_0^x q(t,y)dt + \int_x^y q(x,t)dt$$

What is the function $q(x,y)$?
Solving the integral equation is far from being easy and involves Bessel functions!. We wonder if this integral equation could be solved more easily as in [Jonassen and Knuth, 1978]. The interested reader may see the details in [Jonassen and Knuth, 1978]. Below we summarize their numerical results:

$$a_\infty = 0.15049, \quad b_\infty = 0.19601, \quad c_\infty = 0.35250,$$
$$d_\infty = 0.13731, \quad e_\infty = 0.16366,$$

$$f_\infty = 0.51617, \quad g_\infty = 0.48382.$$

A similar study has been done by Baeza-Yates [Baeza-Yates, 1989] for binary search trees with four nodes. It is really surprising that such a trivial algorithm leads to such non trivial analysis.

2.2 GENERALITIES ON DYNAMIC DATA STRUCTURES

A data type is a specification of the basic operations allowed together with its set of possible restrictions. The following data types are commonly used

Stack: keys are accessed by position, operations are insertion I and deletion D but are restricted to operate on the key positioned first in the structure (the "top" of the stack).

Linear list: keys are accessed by position, operations are insertion I and deletion D without access restrictions (linear lists make it possible to maintain dynamically changing arrays).

Dictionary: keys belonging to a totally ordered set are accessed by value, all four operations I, D, Q^+, Q^- are allowed without any restriction.

Priority queue: keys belonging to a totally ordered set are accessed by value, the basic operations are I and D, deletion D is performed only on the key of minimal value (of "highest priority").

Symbol table: this type is a particular case of dictionary where deletion always operates on the key last inserted in the structure, only positive queries are performed.

A data organization is a machine implementation of a data type. It consists of a data structure which specifies the way objects are internally represented in the machine, together with a collection of algorithms implementing the operations of the data type.

Stacks are almost always implemented by arrays or linked lists .

Linear lists are often implemented by linked lists and arrays.

Dictionaries are usually implemented by sorted or unsorted lists; binary search trees have a faster execution time and several balancing schemes have been proposed: AVL, 2-3 and red-black trees. Other alternatives are h-tables and digital trees.

Priority queues can be represented by any of the search trees used for dictionaries, more interesting are heaps , P-tournamnets, leftist tournaments, binomial tournaments, binary tournamnets and pagodas. One can also use sorted lists, and any of the balanced tree structures.

Symbol tables are special cases of dictionaries, all the known implementations of dictionaries are applicable here.

DEFINITION 3.1 *A schema (or path) is a word*

$$\Omega = O_1 O_2 \ldots O_n \in \{I, D, Q^+, Q^-\}^*$$

such that for all j, $1 \leq j \leq n$:

$$|O_1 O_2 \ldots O_j|_I \geq |O_1 O_2 \ldots O_j|_D.$$

$\{I, D, Q^+, Q^-\}^*$ is the free monoid generated by the alphabet

$$\{I, D, Q^+, Q^-\}$$

$|w|$ is the length of the word w.

A schema is to be interpreted as a sequence of n requests (the keys operated on not being represented).

DEFINITION 3.2 *A structure history is a sequence of the form:*

$$h = O_1(r_1) O_2(r_2) \ldots O_n(r_n)$$

where $\Omega = O_1 O_2 \ldots O_n$ is a schema, and the r_j are integers satisfying: $0 \leq r_j < pos(\alpha_{j-1}(\Omega))$ *and* $\alpha_j(\Omega) = |O_1 O_2 \ldots O_j|_I - |O_1 O_2 \ldots O_j|_D$ *is*

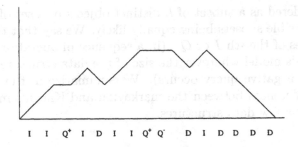

I   I   Q⁺   I   D   I   I   Q⁺   Q⁻   D   I   D   D   D   D

Figure 3.3.   A schema

the size (level) of the structure at step j, pos is a possibility function defined on each request, r_j is the rank (or position) of the key operated upon at step j.

We will only consider schemas and histories with initial and final level 0.

Possibility functions.

Two different models have been considered for defining possibility functions: the markovian model [Flajolet et al., 1981, Françon, 1978] in which possibility functions are linear functions of the size k of the data structure when an allowed operation is performed.

Knuth's model is related to his observation [Knuth, 1977] that deletions may not preserve randomness and is more realistic than the markovian model. The following simple example may be helpful to understand Knuth's fundamental remark.

Consider again the sequence of operations $IIDI$ performed, for example, on a linear list which is initially empty. Let $x < y < z$ be the three keys inserted during the sequence III. x, y and z are deleted with equal probability. Let w be the key inserted by the fourth I. Then all four cases $w < x < y < z$, $x < w < y < z$, $x < y < w < z$, $x < y, z < w$ do occur with equal probability, whatever the key deleted. More generally, let us consider a sequence of operations $O_1 O_2 \dots O_j$ on a dictionary data type, the initial data structure being empty (any data type listed above may be considered). Assume O_j is the ith I or Q^- of the sequence. Let $x_1 < x_2 < \dots < x_{i-1}$ be the keys inserted and negatively searched during the sequence $O_1 O_2 \dots O_{j-1}$, and let w be the ith inserted or negatively searched key. Then all the cases $w < x_1 < x_2 < \dots < x_{i-1}$, $x_1 < w < x_2 < \dots < x_{i-1}$, $\dots$, $x_1 < x_2 < \dots < x_{i-1} < w$ are equallly likely, whatever the deleted keys. Put into combinatorial words: after j operations, whose i are I and Q^-'s (thus $j - i$ are D and Q^+'s), the size of the data structure is $k \leq 2i - j$. The keys of the data structure can be considered as a subset of k distinct objects of a set of size i any of the C_i^k possible subsets being equally likely. We say that the number of possibilities of the ith I or Q^- (in a sequence of operations) is equal to i in Knuth's model whatever the size of the data structure when this insertion (or negative query occurs). We summarize in the two tables below the differences between the markovian and Knuth's models. We consider only a few data structures.

Data type	Npos(I,k)	Npos(D,k)	Npos(Q^+,k)	Npos(Q^-,k)
Dictionary	k+1	k	k	k+1
Prority queue	k+1	1	0	0
Linear list	k+1	k	0	0

Table 3.1.   Possibility functions in the markovian model.

Data type	Npos(I,k)	Npos(D,k)	Npos(Q^+,k)	Npos(Q^-,k)
Dictionary	i	k	k	i
Prority queue	i	1	0	0
Linear list	i	k	0	0

Table 3.2.   Possibility functions in Knuth's model

2.3   ANALYSIS

The markovian model (resp. Knuth's model) has been studied by combinatorial methods in [Flajolet et al., 1981] (resp. [Françon et al., 1990]) and the probabilistic analysis has been done by G. Louchard [Louchard, 1987] and R.S. Maier [Maier, 1991a] (resp. G. Louchard et al. [Louchard et al., 1992]). For both models the probabilistic analysis permits to characterize the limiting distributions of time and storage costs. We explain the main ideas of Louchard's probabilistic analysis through the linear lists in the markovian model. Then we present R.S. Maier's large deviations approach.

2.4 PROBABILISTIC ANALYSIS IN THE MARKOVIAN MODEL

2.4.1 LOUCHARD'S APPROACH

We explain his approach through the linear list example.
Let h be a history, Louchard's idea [Louchard, 1987] is to associate a random variable C^* to any cost function $C(h)$ such that:

$$P(C^* = k) = \frac{card\{h : C(h) = k, h \in \tilde{N}_{2n}\}}{N_{2n}} \qquad (3.7)$$

where $\tilde{N}_{2n}$ is the set of histories of length $2n$ and N_{2n} its cardinality. The linear list is asymptotically equivalent to a classical random walk, $Y(i)$, of length $2n$ from 0 to 0, constrained to remain nonnegative and with weight $\Pi_{i=1}^{2n-1} Y(i)$ corresponding to all histories of a given schema. Because of the weight, the random walk is not asymptotically equivalent to a Brownian excursion9.1

LEMMA 3 *[Louchard, 1987] Let* $Y(i) = \sum_{j=1}^{k} \alpha_j$ *where the* α_j *are independent identically distributed random variables with distribution:* $P(\alpha_j = -1) = P(\alpha_j = +1) = \frac{1}{2}$ *Then, for* $0 < u < 1$ *and* $m \to \infty$,

$$E_0(\frac{Y(m)}{m} \in [u, u + du]) =$$
$$(1 - u^2)^{-\frac{m}{2}} (\frac{1+u}{1-u})^{-m\frac{u}{2}} \frac{\sqrt{m}du}{\sqrt{2\pi(1 - u^2)}} [1 + \frac{\rho(u)}{m} + O(m^{-2})] \qquad (3.8)$$

where

$$\rho(u) = \frac{1 - \frac{3}{2}(1 - u^2)}{\frac{1}{2}(1 - u^2)}. \qquad (3.9)$$

Proof: The probability we seek is equivalent to

$$E_0[\frac{\hat{Y}(m) - \frac{1}{2}m}{m} \in [u, u + du]]$$

with

$$\hat{Y}(i) = \sum_{j=1}^{i} \hat{\xi}_j \quad and \quad P[\hat{\xi}_j = 0] = P[\hat{\xi}_j = 1] = \frac{1}{2}$$

The generating function of $\hat{\xi}_j$ is given by

$$g(z) = \frac{1}{2}(1 + z)$$

Define β by

$$\beta = \frac{1+u}{1-u}$$

The probability we need is now related to the new generating function:

$$\frac{g(\beta z)}{g(\beta)} = \frac{1}{2}(1-u) + \frac{1}{2}(1+u)z \qquad (3.10)$$

with mean $\frac{1}{2}(1+u)$ and variance $V = \frac{1}{4}(1-u^2)$. One can show, with the central limit theorem [Feller, 1970, Louchard, 1987] that for $m \to \infty$

$$E_0[\frac{\hat{Y}(m) - \frac{1}{2}m}{m} \in [u, u+du]] =$$

$$\frac{g(\beta)^m}{\beta^{m\frac{1+u}{2}}\sqrt{2\pi m v^2}}\frac{1}{2}mdu[1 + \frac{\rho(u)}{m} + O(m^{-2})]$$

where

$$\rho(u) = \frac{\mu_4 - 3V^2}{24V^2}H_4(0)$$

μ_4 is the fourth moment of Eq.(3.10) and H_4 is the fourth Hermite polynomial, $H_4(x) = x^4 - 3x^3 + 3$. The result of the lemma is now derived through standard computations. We must now take into account the constraint of nonnegativity for the random walk.

LEMMA 4 *[Louchard, 1987] Let $Y(i) = \sum_{j=1}^{i}\xi_j$ and $\Lambda = min(n \geq 1 : Y(n) \leq 0)$. Define $\tilde{Y}(i) = [Y(i)|\Lambda = 2n]$. Let also: $Y_a^-(i) = [a + Y(i); \Lambda > i]$ be the random walk starting at a and killed at 0. We denote by B an event belonging to the Borel field generated by Y^-, and define*

$$\Pi(v, x, y)dy = E_{nx}[B, \frac{Y_{nx}^-(nv)}{n} \in [y, y+dy]]$$

Let $0 < \epsilon$, $\delta < 1$. The event B for $\tilde{Y}(i)$, $i \in [n\epsilon, n(2 - \delta)]$, has the following asymptotic probability:

$$E_0[\frac{\tilde{Y}(n\epsilon)}{n\epsilon} \in [u_1, u_1+du_1], B, \frac{\tilde{Y}(n(2-\delta))}{n\delta} \in [u_2, u_2+du_2]$$
$$\sim \sqrt{8\pi}(2n)^{3/2}\phi(u_1, \epsilon)\phi(u_2, \delta)\Pi(2 - \epsilon - \delta, \epsilon u_1, \delta u_2)du_1 du_2 \qquad (3.11)$$

where

$$\Phi(u, \epsilon) = \begin{cases} log(\frac{1+u}{1-u})\phi(n\epsilon, u) & if\ u < 1 \\ 2^{-(n\epsilon+1)} & if\ u=1. \end{cases}$$

and $\phi(n\epsilon, u) = \dfrac{\sqrt{n\epsilon}}{\sqrt{2\pi(1-u^2)}}$.

Proof: Assume first that $u < 1$. From the reflection principle [Feller, 1970], we get that, for $m \to \infty$,

$$
\begin{aligned}
E_0[\dfrac{Y_1(m)}{m} \in [u, u+du]] &\sim [\phi(m, u - \dfrac{1}{m}) - \phi(m, u + \dfrac{1}{m})]du \\
&\sim \dfrac{2}{m}\phi(m, u)du \\
&\sim \dfrac{2}{m}[-\dfrac{u}{1-u^2} + \dfrac{1}{2}m log(\dfrac{1+u}{1-u})]\phi(m, u)du \\
&\sim log(\dfrac{1+u}{1-u})\phi(m, u)du
\end{aligned}
$$

By the random walk symmetry, we deduce the first part of Eq.(3.11). Between $n\epsilon$ and $n(2-\delta)$, $\tilde{Y}$ and Y^- are equivalent in probability. The coefficient $\sqrt{8\pi}(2n)^{\frac{3}{2}}$ is a normalizing factor (see [Chung, 1976]). When $u = 1$, the discrete probability related to $Y(m)$ leads to 2^{-m} but $Y(m)$ and m have the same parity, hence the factor $\frac{1}{2}$ in the asymptotic density. The weight defined by Eq.(3.7) and Table 3.1 must now be taken into account. We define the weighted random walk Y^*. We set tentatively

$$\sqrt{n}X_n(v) = Y^*([nv]) - ny(v), v \in [0, 2] \qquad (3.12)$$

where $y(.)$ is a deterministic continuous nonnegative symmetric function with $y(0) = y(2) = 0$ and $X_n(.)$ a random process with asymptotic zero mean (coefficients will be justified in the sequel). Moreover as

$$Y^*(1) = Y^*(2n-1) = 1 \text{ and } Y^* \le n \qquad (3.13)$$

we must have

$$lim_{n \to \infty} ny(\dfrac{1}{n}) = lim_{n \to \infty} ny(2 - \dfrac{1}{n}) = 1, \ ny(.) \le n$$

The constraints we put on $y(.)$ in the sequel are as follows:

$$y \in C^1, \ y(0) = y(2) = 0, \ y(.) \le 1, \ y'(0) = -y'(2) = 1 \qquad (3.14)$$

According to Eq.(3.7), we put on each trajectory of type Eq.(3.13) a total measure which is the product of probability measures as defined by Eq.(3.11) and the weight

$$W = \Pi_{i=1}^{2n-1}Y^*(i)$$

We firstly must find $y(.)$; then establish the stochastic properties of $X_n(.)$ and justify Eq.(3.13). Let us firstly look for an asymptotic formula for W along $ny(.)$. We have:

LEMMA 5 *[Louchard, 1987] Let $y(.)$ satisfy Eq.(3.14). Let*

$$W = exp[\sum_{i=1}^{2n-1} log(ny(\frac{i}{n}))] = exp(Z)(say)$$

and

$$W(j,k) = exp[\sum_{i=j}^{k-1} log(ny(\frac{i}{n}))]$$

Then

$$W \sim exp[2nlogn + n\int_0^2 log(y(v))dv + 2 + \int_{1/n}^{2-1/n} B_1(\{nv\})\frac{y'(v)}{y(v)}dv],$$

$$(3.15)$$

$$W(j.k) \sim exp[(k-1-j)logn + n\int_{j/n}^{k/n} log(y(v))dv - \frac{1}{2}[log(y\frac{k}{n}) - log(y\frac{j}{n})]$$

$$+ \int_{j/n}^{k/n} B_1(\{nv\})\frac{y'(v)}{y(v)}dv$$

$$(3.16)$$

where B_1 is the first Bernoulli polynomial (see [Abramowitz and Stegun, 1965]).
By convention, we say that the convergence condition (CC) is satisfied if the last terms of Eq.(3.15) and Eq.(3.16) converge.

Proof: Let

$$Z = \sum_{i=1}^{2n-2} log(ny(\frac{i}{n}) + log(ny(2 - \frac{1}{n}))$$

The second term tends to 0 by Eq.(3.14). The first term leads, by Euler's summation formula to

$$(2n-2)\log n + \int_1^{2n-1} log(y(\frac{x}{n}))dx - \frac{1}{2}[log(y(\frac{2n-1}{n}) - log(y(\frac{1}{n})) +$$

$$\int_1^{2n-1} B_1(\{x\})f'(x)dx$$

For simplicity we write this expression as $(2n - 2)logn + A + B + C$. We get: $B \to 0$ as $n \to \infty$, and:

$$A = n \int_{1/n}^{2-1/n} log(y(v))dv = n \int_0^2 log(y(v))dv + n \int_0^{1/n} log(y(v))dv$$

$$- n \int_{2-1/n}^2 log(y(v))dv$$

But by Eq.(3.15):

$$-n \int_0^{1/n} log(y(v))dv \sim logn + 1$$

Expression Eq.(3.15) is now straightforward and Eq.(3.16) is similarly proved. We want now to determine $y(.)$.

Let ϵ, δ be small positive constants. Along the trajectory of type Eq.(3.13), for $i \in [2n\epsilon, 2n(1 - \delta)]$, asymptotically, Y^- and Y are identically distributed (the hitting probability of the zero boundary is exponentially small).

Let us assume that y satisfies the conditions Eq.(3.14). When we let ϵ, $\delta \to 0$, it follows from Eq.(3.11) that the total probability measure on Eq.(3.13) is asymptotically (as $n \to \infty$) the same measure as the limiting one (ϵ, $\delta \to 0$) deduced from Eq.(3.8) for $i \in [2n\epsilon, 2n(1 - \delta)]$, multiplied by

$$\frac{1}{4}(2n)^{3/2}2^{-n(\epsilon+\delta)} \tag{3.17}$$

We now define x_1, $x_2 \in [\epsilon, 2 - \delta]$, $x_2 = x_1 + \Delta$ (with $\Delta > 0$), $y_1 = y(x_1)$, $y_2 = y(x_2)$. The transition probability from nx_1, ny_1 to $nx_2, n(y_2 + dy_2)$, as given by Eq.(3.8), leads, if we neglect $\rho(u)$ to

$$(1 - u^2)^{-n\Delta/2}(\frac{1+u}{1-u})^{-n\Delta/2}\frac{\sqrt{n}dy_2}{\sqrt{2\pi\Delta(1 - u^2)}} \tag{3.18}$$

with $u = \frac{n(y_2-y_1)}{n\Delta}$.

The dominant term in the *log* of this density is given by

$$\frac{-1}{2}n\Delta[log(1 - u^2) + ulog(\frac{1+u}{1-u})].$$

Now as $\Delta \to 0$ (in a sense to be precised later on), $u \to y'(x_1)$. The dominant term in the log of the total asymptotic probability measure along $ny(.)$ is now derived as

$$\frac{-1}{2}n \int_{\epsilon}^{2-\delta} [log(1 - y'(v)^2) + y'(v)log(\frac{1 + y'(v)}{1 - y'(v)})dv. \qquad (3.19)$$

Three problems remain to be solved:

- We must take into account, in Eq.(3.18), the factor $\frac{\sqrt{n}}{\sqrt{2\pi\Delta(1-u^2)}}$.

- The factor $2^{-n\epsilon}$ in Eq.(3.17) must be considered. It can easily be checked that, if the function $y(.)$ satisfies the condition

$$y'(v) = 1 + Kv^j + 0(v^j), \quad j \geq 1 \qquad (3.20)$$

then

$$2^{-n\epsilon} \sim_{\epsilon \to 0} exp[\frac{-1}{2}n \int_0^{\epsilon} [log(1 - (y')^2) + y'log(\frac{1 + y'}{1 - y'})dv].$$

The integral in Eq.(3.18) can thus be extended to $[0, 2]$.

- We must check that we can neglect the contribution from $\rho(u)$ in Eq.(3.8). This can be done by carefully coupling Δ with n. To ease the analysis, let $\epsilon = \delta = \Delta = 2/k$.

By Eq.(3.9), we must add to Eq.(3.18) a term which is

$$\sim \frac{1}{n\Delta} \sum_{i=1}^{k-2} \rho[y'(\frac{2i}{k})]$$

By Euler's summation formula, this is equivalent to

$$\frac{1}{n\Delta} \int_1^{k-1} \rho[y'(\frac{2x}{k})]dx - \frac{1}{2n\Delta}[\rho[y'(2 - \Delta)] - \rho[y'(\Delta)]]$$

$$+ \frac{1}{n\Delta} \int_1^{k-1} B_1(x)\delta_x \rho[y'(\frac{2x}{k})]dx = term(a) + term(b) + term(c)$$

Let us assume that Eq.(3.20) is satisfied. Then by Eq.(3.9)

$$\rho[y'(v)] = \frac{K}{v^j} + O(\frac{1}{v^{j-1}}) \ for \ some \ K.$$

Term(a) becomes:

$$\frac{1}{n\Delta^2}\int_{\Delta}^{2-\Delta}\rho[y'(v)]dv$$

and $term(a) \to 0$ if $n\Delta^{j+1} \to_{n\to\infty} \infty$

$Term(b) \sim \frac{1}{n\Delta^j} \to 0$, if $n\Delta^j \to \infty$

$Term(c) \to 0$, if $\frac{1}{n\Delta^{j+1}} \to 0$

The contribution from $\rho(.)$ can thus be neglected by letting $n \to \infty$, $\Delta \to 0$ with $n\Delta^{j+1} \to \infty$.

Note that, with this last condition, $term(b) \to 0$, even if the measure in Eq.(3.19) is computed on $[\epsilon, v]$, $v < 2$.

Summarizing our results, we derive the following lemma.

LEMMA 6 *[Louchard, 1987] The dominant term in the log of the total asymptotic probability measure along $ny(.)$, between v_1 and v_2, is given, if Eq.(3.20) is satisfied, by*

$$-\frac{1}{2}n\int_{v_1}^{v_2}[log(1-y'(v)^2)+y'(v)log(\frac{1+y'(v)}{1-y'(v)}]dv. \qquad (3.21)$$

Turning now to the weight, we see that, if the condition (CC) of Lemma 5 is satisfied, the dominant term in the log of the weight along $ny(.)$, between v_1 and v_2 is given, is satisfied, by

$$(v_2-v_1)nlogn + n\int_{v_1}^{v_2}log(y(v))dv \qquad (3.22)$$

Collecting the results of Eq.(3.21) and Eq.(3.22), we finally obtain the next lemma.

LEMMA 7 *[Louchard, 1987] The dominant term in the log of the asymptotic total measure along $ny(.)$, between v_1 and v_2, is given, if the conditions of the previous lemmas are satisfied (with $v_1 = j/n$, $v_2 = k/n$, by:*

$$(v_2-v_1)nlogn + n\int_{v_1}^{v_2}[-\frac{1}{2}log(1-(y'(v))^2)-\frac{1}{2}y'(v)log(\frac{1+y'(v)}{1-y'(v)})$$
$$+ log(y(v))]dv = (v_2-v_1)nlogn + n\int_{v_1}^{v_2}f(y,y')dv.$$
$$(3.23)$$

Expressions Eq.(3.21) and Eq.(3.22) are balanced in the sense that their variable (in $y(.)$) part is linear in n. It is easily seen that any other function $h(n)y(v)$, with $h(n) = 0(n)$ would lower the first term in Eq.(3.15) without increasing the probability measure Eq.(3.21) beyond $O(1)$. The total measure in Eq.(3.23) would only be smaller. On the other side, $Y^* \leq n$, and we cannot go beyond $h(n) = n$.

The determination of $y(.)$ is finally solved by the following theorem.

THEOREM 3.3 *[Louchard, 1987] For linear lists*

$$y(v) = \frac{1}{C_1} \sin(C_1(v - C_2)), \tag{3.24}$$

where C_1 and C_2 are some constants. The boundary conditions, $y(0) = y(2) = 0$, lead to $C_1 = \frac{1}{2}\pi$, $C_2 = 0$. The other conditions are obviously satisfied.

Proof: The Laplace method leads to the following question: find the function $y(.)$ maximizing Eq.(3.23). This is a classical variational problem.

As f in Eq.(3.23) does not depend on v, we know that the associated Euler-Lagrange equation possesses an integral of the form

$$f - y' f_{y'} = L$$

where L is a constant, which easily leads to Eq.(3.24).

Now we analyze the distribution of X_n, $x \in [0, 2]$.

Set $\theta = X_n(x)/\sqrt{n}$. Assuming Eq.(3.12) to be valid, it is clear from Lemma 7 and 3.3 that the log of the asymptotic total measure along any trajectory from 0 to $n[y(x) + \theta]$ must be computed on $n[z(v, y(x) + \theta, x) + \sqrt{n} + \sqrt{n}\chi(x)]$, where, by Theorem 3.3,

$$z(v, u, x) = (1/C)sin(Cv) \text{ with } C(u, x) \text{ such that}$$

$$z(x, u, x) = \frac{1}{C}sin(Cx) = u$$

and $\chi(.) \in C^1$ with $\chi(0) = \chi(x) = 0$.

LEMMA 8 *[Louchard, 1987] The dominant term in the asymptotic total measure along $n[z(v, y + \theta, x) + \eta(v)]$, $v \in [0, x]$ and along $z(2 - v, y + \theta, 2 - x)$, $v \in [x, 2]$ is given by:*

$$2nlogn + n \int_0^2 f(y, y')dv - \frac{n\theta^2\pi^2}{4\gamma(x)\gamma(2 - x)}$$

with

$$\gamma(v) = \frac{1}{2}\pi \cos(\frac{1}{2}\pi v)v - sin(\frac{1}{2}\pi v).$$

Proof: See [Louchard, 1987].

We can now draw two important conclusions from Lemma 8.

- The second part of Eq(3.18) is now justified: θ is clearly a term of order $1/\sqrt{n}$.

- Normalizing the result of Lemma 8, in the sense of Eq.(3.7), the limit of $X_n(x)$ is obviously a gaussian variable with mean 0 and variance

$$V^* = 2\gamma(x)\gamma(2-x)/\pi^2$$

Lemma 8 is not sufficient to draw definitive conclusions on the process $X(.)$. We must compute the covariance of $X(.)$ and check the tightness of the sequence $X(.)$ to obtain weak convergence.
The covariance problem is solved by the following lemma.

LEMMA 9 *[Louchard, 1987] The covariance of $X_n(.)$ is asymptotically given by $(x_1 \leq x_2)$:*

$$E(X_n(x_1)X_n(x_2)) \sim 2\gamma(x)\gamma(2-x)/\pi^2 = C^*_{12}$$

The limiting process is markovian.

Proof: The proof of this lemma is fairly technical. See [Louchard, 1987] for details.
Collecting our results, we finally state this theorem.

THEOREM 3.4 *[Louchard, 1987] Let*

$$\gamma(v) = \frac{1}{2}\pi \cos(\frac{1}{2}\pi v)v - sin(\frac{1}{2}\pi v),$$

for linear lists:

$$\frac{Y^*([nv]) - n(2/\pi)sin(\frac{1}{2}\pi v)}{\sqrt{n}} \Rightarrow_{n\to\infty} X(v)$$

where $X(.)$ is a markovian gaussian process with mean 0 and covariance $C^*_{12} = 2\gamma(x)\gamma(2-x)/\pi^2$ $X(.)$ can be written as

$$X(v) = \frac{\sqrt{2}\gamma(2-v)}{\pi}B(\frac{\gamma(v)}{\gamma(2-v)})$$

where $B(.)$ is a Browniam motion.

From Theorem 3.4 we derive the cost functions:
a) Storage cost function

THEOREM 3.5 *Let Σ_{LL}^* be the storage cost function of the linear list (LL).*

$$\frac{\Sigma_{LL}^* - n^2 \nu_1}{(n^3 \nu_2)^{1/2}} \sim_{n \to \infty} N(0, 1)$$

where $\nu_1 = \frac{8}{\pi^2}$, $\nu_2 = (128/\pi^2 - 12)/\pi^2$.

b) Time cost function

THEOREM 3.6 *Let τ_{LL}^* be the time cost function of the linear list.*

$$\frac{\tau_{LL}^* - n^2 \nu_3}{(n^3 \nu_4)^{1/2}} \sim_{n \to \infty} N(0, 1)$$

where $\nu_3 = \frac{4}{\pi^2}$, $\nu_4 = 32/\pi^2 - 8/3$.

A similar study can be done for the other dynamic data structures under investigation: priority queues and dictionaries. We summarize Louchard's results below:

THEOREM 3.7 *For priority queues (PQ), the storage cost function Σ_{PQ}^* behaves as follows:*

$$\frac{\Sigma_{PQ}^* - \frac{2}{3}n^2}{(\frac{8}{45}n^3)^{1/2}} \sim_{n \to \infty} N(0, 1)$$

The time cost function behaves as:

$$\frac{\tau_{PQ}^* - \frac{1}{3}n^2}{(\frac{2}{45}n^3)^{1/2}} \sim_{n \to \infty} N(0, 1)$$

if the list is unsorted and as

$$\frac{\tau_{PQ}^* - \frac{1}{6}n^2}{(\frac{1}{45}n^3)^{1/2}} \sim_{n \to \infty} N(0, 1)$$

if the list is sorted.

THEOREM 3.8 *For dictionaries (D), the storage cost function Σ_D^* behaves as follows*

$$\frac{\Sigma_D^* - \frac{2}{3}n^2}{(\frac{4}{45}n^3)^{1/2}} \sim_{n \to \infty} N(0, 1)$$

The time cost function behaves as

$$\frac{\tau_D^* - \frac{1}{3}n^2}{(\frac{2}{45}n^3)^{1/2}} \sim_{n \to \infty} N(0,1).$$

REMARK 12 *The probabibistic analysis of dynamic data structures in Knuth's model is more intricate and has been done in [Louchard et al., 1992].*

2.4.2   A LARGE DEVIATIONS APPROACH

R.S. Maier's [Maier, 1991a] analysis leads to the same conclusions as above but by application of the theory of large deviations (cf. page 25). We show below how this approach works.

First we fix some notations.

Let α_j be the number of records contained in a data structure after j operations and $t^{X,Y}(O, \alpha, r)$ the time needed to perform operation O on a structure of size α if the key used has rank r. The superscripts X and Y denote datatype (LL, PQ, D) (i.e. Linear List, Priority Queue, Dictionary) and implementation (SL, UL) (i.e. Sorted List, Unsorted List) respectively.

Denote by r_j the relative rank of the key used in the jth operation, and by O_j the operation itself. Then the inequalities

$$0 \le r_j \le m(\alpha_{j-1}, O_j) - 1$$

in which

$$m(\alpha, O) = \begin{cases} \alpha & O = D, Q^+ \\ \alpha + 1 & O = I, Q^- \\ 1 & O = D_{min} \end{cases} \qquad (3.25)$$

D_{min} is 'delete min' operation, it acts only on the key of minimum rank.

Define the normalized data structure size, as a function of time, to be

$$x(t) = \frac{1}{n}\alpha_{[nt]}, t \in [0,1]$$

To leading order in n, the normalized integrated space cost

$$\frac{S}{n^2} \sim \int_0^1 x(t)dt$$

So the asymptotics of the random process $x(t)$ will determine those of $\frac{S}{n^2}$. The probability measure on the space of $n \to \infty$ continuous limiting paths $\{x(t)\}_{0 \le t \le 1}$ can be obtained from the joint probability distribution of the $\{\alpha_j\}_{j=0}^n$. The assumed equiprobability of operation words implies that an allowed size history α_j has relative probability

$$\Pi_{j=1}^n m(\alpha_{j-1}, O_j) \tag{3.26}$$

To leading order in n this is independent of the operations O_j, equalling

$$\left\{ \begin{array}{ll} \Pi_{j=1}^n(\alpha_j + 1), & X \in \{LL, D\} \\ \Pi_{j=1}^n \sqrt{(\alpha_j + 1)}, & X \in \{PQ\} \end{array} \right. \tag{3.27}$$

by Eq.(3.25) The weighting factors Eq.(3.27), which can be written as

$$e^{a^X \sum_{j=1}^n log(\alpha_j + 1)}$$

with the datatype-dependent constant

$$a^X = \left\{ \begin{array}{ll} 1, & X \in \{LL, D\} \\ \frac{1}{2}, & X \in \{PQ\} \end{array} \right.$$

affect the relative probabilities of limiting paths $\{x(t)_{0 \le t \le 1}\}$ by weighting them by

$$e^{na^X \int_0^1 log(x(t))dt} \tag{3.28}$$

However, to obtain a complete expression for the probability measure on the space of normalized paths, one must determine what it would have been in the absence of Eq.(3.26) or Eq.(3.28). That is to say, if α_j were an unbiased random walk built up from the operations I, D, Q^+, Q^-, what probabilty measure on the paths $\{x(t)_{0 \le t \le 1}\}$ would be obtained by taking the $n \to \infty$? A similar limit, though without the constraint $\alpha_j \ge 0$, $\forall j$ is familiar from probability theory as the definition of the Wiener process [Billingsley, 1968]. It concerns the scaled random walk

$$w(t) = \sqrt{n} \alpha_{[nt]}$$

The underlying joint distribution of our $\{a_j\}_{j=0}^n$ is

$$\left\{ \begin{array}{ll} \delta_{\alpha_n,0} \delta_{\alpha_0,0} \Pi_{j=0}^n (\delta_{\alpha_j,\alpha_{j-1}-1}) + 2\delta_{\alpha_j,\alpha_{j-1}} + \delta_{\alpha_j,\alpha_{j-1}+1}/4 & X = D \\ \delta_{\alpha_n,0} \delta_{\alpha_0,0} \Pi_{j=0}^n (\delta_{\alpha_j,\alpha_{j-1}-1}) + \delta_{\alpha_j,\alpha_{j-1}+1}/2 & X \in \{LL, PQ\} \end{array} \right.$$

on account of the equiprobability of I, D, Q^+, Q^-.
The two cases $X = D$ and $X \in \{LL, PQ\}$ are said to have diffusion constants

$$D^X = \begin{cases} 1/2, & X = D \\ 1, & X \in \{LL, PQ\} \end{cases}$$

since if the constraint $\alpha_n = 0$ were dropped, D^X would be the rate at which α_j diffuses as j increases. The unnormalized $n \to \infty$ limiting measure on paths $w(.)$ is then given by the Feynman-Kac formula [Ito and McKean, 1974]

$$e^{-\int_0^1 \frac{|\dot{w}(t)|^2}{2D^X}} \mathcal{D}w \tag{3.29}$$

Here $\mathcal{D}w$ signifies a formal infinite-dimensional measure on the space of paths $\{w(t)\}_{0 \le t \le 1}$. For finite n

$$w(t) = n^{1/2}x(t) \tag{3.30}$$

so one might think that the analogous formula for the limiting measure on paths $\{x(t)_{0 \le t \le 1}$ would be

$$e^{-n\int_0^1 \{\frac{|\dot{x}(t)|^2}{2D^X}\}dt} dx \tag{3.31}$$

Equation Eq.(3.31) contains n explicitly, and cannot be interpreted as a straightforward $n \to \infty$ limit. It could, however, be interpreted as a large deviation principle, governing the large n asymptotics of the measure. However Eq.(3.31) is incorrect as it stands.
R. Maier [Maier, 1991a] has shown that the correct expression is the more complicated large deviation principle

$$e^{-n\int_0^1 \{\frac{[(1+\dot{x}(t))\log(1+\dot{x}(t))+(1-\dot{x}(t))\log(1-\dot{x}(t))]}{2D^X}\}dt} dx \tag{3.32}$$

in which the integrand is understood to equal $+\infty$ if $\dot{x}(t) \notin (-1,1)$.
Given the validity of Eq.(3.32) one can combine Eq.(3.28) with it to obtain the limiting measure on paths $\{w(t)\}_{0 \le t \le 1}$

$$e^{-n\int_0^1 \{\frac{[(1+\dot{x}(t))\log(1+\dot{x}(t))+(1-\dot{x}(t))\log(1-\dot{x}(t))]}{2D^X - a^X \log x(t)}\}dt} dx \tag{3.33}$$

which we write as

$$e^{-n\int_0^1 L(x(t),\dot{x}(t))dt} dx \tag{3.34}$$

where

$$L(x,\dot{x}) = \frac{[(1+\dot{x})\log(1+\dot{x}) + (1-\dot{x})\log(1-\dot{x})]}{2D^X} - a^X \log x \tag{3.35}$$

provided $\dot{x} \in (-1, 1)$; otherwise, $L(x, \dot{x}) = \infty$.

Equations Eq.(3.33) and Eq.(3.34) are understood to apply to differentiable paths $\{x(t)\}_{0 \le t \le 1}$ that satisfy the constraints

$$x(t) \ge 0, \ t \in [0, 1], \ x(0) = 0, \ x(1) = 0.$$

They are consequences of the constraints on the operation words. The relative probabilities of the normalized data structure sizes are as $n \to \infty$ given by Eq.(3.34). It implies that for finite n the most likely sizes, as functions of time, are those for which the integral

$$\int_0^1 L(x(t), \dot{x}(t))dt \tag{3.36}$$

is minimized, and that as $n \to \infty$ the model gives negligible weight to all others. The problem of minimizing such a functional of paths as Eq.(3.36) is solved by the calculus of variations. The extremizing path $x_X^*(.)$ will satisfy the Euler-Lagrange equation

$$\frac{d}{dt}\frac{dL}{d.x} - \frac{dL}{dx} = 0 \tag{3.37}$$

So by Eq.(3.35) $x_X^*(.)$ satisfies

$$\frac{\ddot{x}_X^*}{1 - (\dot{x}_X^*)^2} = \frac{D^X \alpha^X}{x_X^*}$$

a nonlinear differential equation that depends on the choice of datatype only through the product

$$D^X \alpha^X = \begin{cases} \frac{1}{2}, & X \in \{D, PQ\} \\ 1, & X \in \{LL\} \end{cases}$$

Its solutions are displayed below

X	$(x_X^*)(t)$
LL	$\frac{1}{\pi}\sin \pi t$
D	$t - t^2$
PQ	$t - t^2$

Since $x(t) = x_X^*(t) + \frac{1}{\sqrt{n}}z(t)$, one may expand $L(x(t), \dot{x}(t))$ in a formal Taylor series about $x(t) = x_X^*(t)$. Substitution of this series into

Eq.(3.34) yields

$$exp - \{\frac{1}{2}\int_{t=0}^{1}\{\frac{\dot{z}^2(t)}{D^X[1-(x_X^*(t)^2]} + \frac{\alpha^X z^2(t)}{(x_X^*(t))^2}\}dt + O(n^{-\frac{1}{2}})\}dz \quad (3.38)$$

or equivalently, by integration by parts,

$$exp - \{\frac{1}{2}\int_{t=0}^{1}\{z(t)(\frac{-d}{dt}\frac{1}{D^X(1-(x_X^*(t))^2)}\frac{d}{dt} + \frac{\alpha^X}{(x_X^*(t))^2})z(t)\}dt$$

$$+ O(n^{-\frac{1}{2}})\}dz$$

$$(3.39)$$

In Eq.(3.38) and Eq.(3.39) we have dropped a constant factor arising from the constant term in the Taylor series. The expression Eq.(3.39) implies that the $\{z(t)\}_{0\leq t\leq 1}$ are asymptotically jointly gaussian, with inverse covariance matrix equal to

$$-\frac{d}{dt}\frac{1}{D^X(1-(x_X^*(t))^2}\frac{d}{dt} + \frac{\alpha^X}{(x_X^*(t))^2} \quad (3.40)$$

The $O(n^{-\frac{1}{2}})$ terms in Eq.(3.39), which are of cubic and higher order in z and $\dot{z}$, yield corrections to the leading gaussian behavior.

The limiting covariance of the $\{z(t)\}_{0\leq t\leq 1}$ may be computed by inverting Eq.(3.40) on the space of differentiable functions $z(.)$ obeying boundary conditions at $t = 0$ and $t = 1$. That is to say, the limiting scaled covariances

$$C^X(t_1, t_2) = lim_{n\to\infty}Ez(t_1)z(t_2)$$
$$= lim_{n\to\infty}nCov(x(t_1), x(t_2))$$

when regarded as functions of t_1 will satisfy the differential equation

$$-\frac{d}{dt_1}\frac{1}{D^X(1-x_X^*(t_1)^2)}\frac{dC}{dt_1} + \frac{\alpha^X}{x_X^*(t_1)^2}C = \delta(t_1 - t_2)$$

in which $\delta(.)$ is the Dirac delta function. This equation can be solved by inspection in the cases $X \in \{LL, D, PQ\}$. The asymptotics of the variances of $\frac{S^X}{n^2}$ can be computed from the C^X, as

$$var(\frac{S^X}{n^2}) \sim E(\int_{t_1=0}^{1}x(t_1)dt_1\int_{t_2=0}^{1}x(t_2)dt_2) - (E\int_{t_1=0}^{1}x(t_1)dt)^2$$

$$= \int_{t_1=0}^{1}\int_{t_2=0}^{1}cov(x(t_1), x(t_2))dt_2dt_1$$

$$\sim \frac{1}{n}\int_{t_1=0}^{1}\int_{t_2=0}^{1}C^X(t_1, t^2)dt_2dt_1$$

The asymptotic standard deviation $\sigma(S^X/n^2)$ is tabulated below for the three datatypes.

X	$E(S^X/n^2)$	$\sigma(S^X/n^2)$
LL	$2\pi^{-2}$	$(16\pi^{-4} - \frac{3\pi^{-2}}{2})^{\frac{1}{2}} n^{-\frac{1}{2}}$
D	$\frac{1}{6}$	$(\frac{1}{6\sqrt{5}}) n^{-\frac{1}{2}}$
PQ	$\frac{1}{6}$	$(\frac{1}{3\sqrt{5}}) n^{-\frac{1}{2}}$

2.5   DATA STRUCTURES' MAXIMA

This section concerns the maximum size reached by a dynamic data structure over a long period of time. The notion of "maximum" is basic to resource preallocation. See also E. Castillo's book [Castillo, 1988] for applications of the extreme value theory in engineering.

We restrict our presentation to list structures and present the probabilistic analysis done in [Louchard et al., 1997]. In this section, we look precisely at the distribution of the maximum size. Assume that the operations happen at times $1, 2, \ldots, n$. The technique is based on the observation that the process can be decomposed into two simple components: let $S(t)$ denote the size of the data structure at time t ($1 \leq t \leq n$). Then $S(t) = n(\tilde{y} + Z(t))$, where $\tilde{y}$, the *average* size of the structure at time t, is a fairly simple curve (for instance a concave parabola in our first subsection), and $Z(t)$ is a (small) gaussian process.

2.5.1   DANIELS' FUNDAMENTAL RESULT

All the results in this section are based on a general theorem by Daniels [Daniels, 1989]. We want information about $Max_{t \in [0\ldots1]} Y(t)$, where Y is a certain random process. Assume that $Y(t)$ can be written as

$$Y(t) = \tilde{z}(t) + Z(t),$$

where $\tilde{z}(t)$ is a certain deterministic curve and $Z(t)$ is a random gaussian process, of covariance $C(s,t)$. Note that $\tilde{z}(t)$ is *not* random. Let M be the maximum of $Y(t)$ for $t \in [0\ldots1]$, and t^* be the first time at which the maximum M is reached. At that time, $Z(t^*) = M - \tilde{z}(t^*)$, and that is the first time that Z reaches that value. Thus we can look for the hitting time of $Z(t)$ to the absorbing boundary $M - \tilde{z}(t)$. Near that crossing point, $Z(t)$ locally behaves like a Brownian motion 9.1(or a variant of it, such as a Brownian bridge [Durbin, 1985]. It is also known that the hitting time and place densities for a Brownian bridge can be deduced from the hitting time density for a Brownian motion (see for instance

[Louchard et al., 1992] for a constant boundary and [Csaki et al., 1987] for a general proof).

Suppose that we are looking at the maximum size of a data structure over time interval [0...1] when some parameter n goes to infinity (for instance, n might be the number of operations). We will assume that $\tilde{z}(t)$ satisfies

$$\tilde{z}(t) = \sqrt{n}z(t)$$

where $z(t)$ is independent of n (that assumption will be true for all the applications in this section). Moreover, assume that $z(t)$ has a unique maximum, reached at time $\bar{t}$, with $z(\bar{t}) = 0$ (up to doing a translation we can always assume that).

Daniels has matched the local behaviour of $C(s,t)$ with the Brownian bridge covariance near $\bar{t}$ [Daniels, 1989]. From [Daniels, 1989] and Daniels and Skyrme [Daniels and Skyrme, 1985], we can deduce information about the maximum M and the time t^* when it is reached.

Notations

Let A and B be the constants defined by

$$A = [\partial_s C]_{\bar{t}} + |[\partial_t C]_{\bar{t}}| \text{and } B = (-z''(\bar{t}))^{-1/3}.$$

Let

$$G(x) = \frac{2^{-1/3}}{2\pi i} \int_{-i\infty}^{i\infty} e^{sx} \frac{ds}{A_i(-2^{1/3}s)}, \qquad (3.41)$$

where A_i is the classical Airy function, and let λ be the universal constant defined by

$$\lambda = \int_{-\infty}^{\infty} [e^{x^3/6}G(x) - max(x,0)]dx \sim 0.99615.$$

Let

$$u = n^{1/3}A^{-1/3}B^{-2}(t^* - \bar{t}).$$

THEOREM 3.9 *With the above assumptions and notations, if* $[\partial_t C(s,t)]_{\bar{t}} \le 0$ *and* $[\partial_s C(s,t)]_{\bar{t}} > 0$, *then we have:*

1. *M is asymptotically gaussian with mean and variance*

$$\begin{cases} E(M) &= \lambda B n^{-1/6} A^{2/3} + O(n^{-1/3}). \\ \sigma^2(M) &= C(\bar{t},\bar{t}) + O(n^{-1/3}). \end{cases}$$

2. *The conditional maximum $M|t^*$ is asymptotically gaussian with mean and variance*

$$\begin{cases} E(M|t^*) &= n^{-1/6}A^{-1/3}B[[\partial_s C]_{\bar{t}}\frac{G'(-u)}{G(-u)} + |[\partial_t C]_{\bar{t}}|\frac{G'(u)}{G(u)}] + O(n^{-1/3}). \\ \sigma^2(M|t^*) &= C(\bar{t},\bar{t}) + O(n^{-1/3}). \end{cases}$$

3. *The joined density of M and t* is given by*

$$\phi(M,t^*) = \frac{2}{\sqrt{2\pi C(\bar{t},\bar{t})}} e^{-M^2/2C(\bar{t},\bar{t})} \left\{ G(t)G(-t) + n^{-1/6} BA^{-1/3} \frac{M}{C(\bar{t},\bar{t})} \right.$$

$$\left. [[\partial_s C]_{\bar{t}} G(t)G'(-t) + |[\partial_t C]|_{\bar{t}} G(-t)G'(t)] + O(n^{-1/3}) \right\}.$$

4. *u has density* $2G(u)G(-u)[1 + O(n^{-1/3})]$.

Proof: We present below the direct new proof given in [Louchard et al., 1997]. Daniel's results [Daniels and Skyrme, 1985, Daniels, 1989] are based on the following hitting time density for a Brownian motion (9.1) $X(t)$.
Let $w(t) = m + \sqrt{n} f(t)$, with $f(t_o) = f'(t_o) = 0$. Daniels has obtained the density (see Daniels and Skyrme [Daniels and Skyrme, 1985]):

$$g(t)dt = P[\min(t' : X(t') = w(t')) \in dt] = \frac{e^{-\frac{[w(t)]^2}{2t}}}{\sqrt{2\pi t}} \mu(t)\, dt. \qquad (3.42)$$

where

$$\mu(t) = n^{-1/3}(2\beta)^{1/3} F\left\{ n^{1/3}[2\beta]^{2/3}(\bar{t} - t) \right\} \left(1 + O(n^{-1/3}) \right)$$

with F given by Eq.(3.41), $\beta := f''(\bar{t})/2$ and $\bar{t}$ is such that $h'(\bar{t}) = h(\bar{t})/\bar{t}$.

(i) We will firstly extract all dominant terms from Eq.(3.42). As in Daniels and Skyrme [Daniels and Skyrme, 1985], set

$$x = n^{1/3}(2\beta)^{2/3}(t_0 - t) = O(1)$$

They have computed

$$\bar{t} - t_o = n^{-1/2} \frac{m}{t_0 2\beta} + O(n^{-2/3})$$

so

$$\bar{t} - t = \frac{xn^{-1/3}}{(2\beta)^{2/3}} + \frac{n^{-1/2}m}{2\beta t_0} + O(n^{-2/3})$$

Expanding $G(x)$ as given by Eq.(3.41), we obtain

$$G(x) = \frac{1}{2^{2/3}} \sum_{k=o}^{\infty} \frac{e^{-\lambda_k x/2^{1/3}}}{A_i'(\lambda_k)}$$

where λ_k are the zeros of $A_i(x)$. A detailed expansion of Eq.(3.42) now leads to

$$g(t) = \frac{(2\beta)^{1/3}}{\sqrt{2\pi t_o}\, 2^{2/3}} \exp\left[\frac{-m^2}{2t_0} + \frac{x^3}{6}\right]$$

$$\times \sum_{k=o}^{\infty} \frac{\exp\left[\frac{-\lambda_k}{2^{1/3}}\left(x + \frac{m}{t_o(2\beta)^{1/3}n^{1/6}}\right)\right]}{A_i'(\lambda_k)} \left(1 + O(n^{-1/3})\right) \quad (3.43)$$

(ii) Such a simple formula as Eq.(3.43) should be explained in direct hitting time density for Brownian motion . This will be done with a technique introduced by Salminen [Salminen, 1988]. Let us first remark that, to derive Eq.(3.43), it is enough to limit ourselves to a second-order boundary (error is within $O(n^{-1/3})$). So we can simply use

$$w(t) = m + \sqrt{n}\beta(t - t_o)^2$$

with $\bar{x} = w(0) = m + \beta\sqrt{n}\,t_o^2$. Our hitting problem can now be transformed into

$$g(t)dt = P_{\bar{x}}[\min(t' : X(t') = h(t')) \in dt]$$

with $h(t) = -\sqrt{n}\,\beta(t - t_o)^2 + \sqrt{n}\,\beta t_o^2$ such that $h(0) = 0$.
This also gives $h'(t) = -2\sqrt{n}\,\beta(t - t_o); h'' = -2\beta\sqrt{n}$.
By Salminen [Salminen, 1988] (2.6) and (3.9), we now obtain

$$g(t) = 2\left(\frac{\beta\sqrt{n}}{2}\right)^{2/3} \exp\left[h'(0)\bar{x} - \frac{1}{2}\int_o^t [h'(s)]^2 ds\right]$$

$$\sum_{k=o}^{\infty} e^{\lambda_k(2\beta^2 n)^{1/3}t} \frac{A_i\left[\lambda_k + 2(\frac{\beta\sqrt{n}}{2})^{1/3}\bar{x}\right]}{A'(\lambda_k)} \quad (3.44)$$

But $h'(0) = 2\sqrt{n}\,\beta t_o$ and

$$-\frac{1}{2}\int_o^t [h'(s)]^2 ds = -\int_o^t 2n\beta^2(s - t_o)^2 ds$$

$$= -2n\beta^2\left[\frac{(t - t_o)^3}{3} + \frac{t_0^3}{3}\right] = \frac{x^3}{6} - \frac{2n\beta^2 t_0^3}{3}$$

By Abramowitz and Stegun [Abramowitz and Stegun, 1965] (10.4.59) we know that, for large z,

$$A_i(z) \sim \frac{1}{2}\pi^{-1/2}z^{-1/4}e^{-\xi}(1 + O(\frac{1}{\xi}))$$

with $\xi := \frac{2}{3}z^{3/2}$.

From Eq.(3.44), we must use

$$z = \lambda_k + 2^{2/3}n^{2/3}\beta^{4/3}t_o^2 + 2^{2/3}\beta^{1/3}mn^{1/6}$$

so that

$$\xi = \frac{4}{3}n\beta^2 t_o^3 \left[1 + \frac{3}{2}\frac{m}{\beta t_o^2\sqrt{n}} + \frac{3}{8}\frac{m^2}{\beta^2 t_o^4 n} + \frac{3}{2}\frac{\lambda_k}{2^{2/3} n^{2/3}\beta^{4/3} t_o^2} \right.$$
$$\left. + \frac{3}{4}\frac{m\lambda_k}{\beta^{7/3} t_o^4 2^{2/3} n^{7/6}} \right] \left(1 + O(\frac{1}{n^{1/3}}) \right)$$

Also, we deduce

$$\lambda_k(2\beta^2 n)^{1/3}t = \lambda_k(2\beta^2 n)^{1/3}t_o - \frac{\lambda_k x}{2^{1/3}}$$

$$h'(0)\bar{x} = 2\beta\sqrt{n}\,t_o\bar{x} = 2\beta\sqrt{n}\,mt_o + 2n\beta^2 t_o^3$$

Identification of Eq.(3.44) with Eq.(3.43) is now routine.

2.5.2   MAXIMA OF THE PRIORITY QUEUE IN THE MARKOVIAN MODEL

Here we study the example of priority file histories. Other structures such as linear lists or dictionaries could be analyzed with similar techniques. Remember that a priority list is a data structure on which no queries are performed: only insertions and deletions occur, and moreover, deletions happen only for the minimum. Thus, when a new item is inserted in a priority list of size k, there are $(k+1)$ intervals defined by the elements already present, and to which the new item may belong, but when an item is deleted, there is only one possible choice. This is reflected in the weights of the paths.

Assume that there are $2n$ operations performed during the history, n insertions and n deletions. Let $Y(t)$ be the size of the data structure after the $\lfloor nt \rfloor^{th}$ operations ($0 \le t \le 2$). From [Louchard, 1987], we know that

$$\forall t, \quad \frac{Y(t) - \frac{1}{2}nt(2-t)}{\sqrt{n}} \Rightarrow X(t) \quad \text{when } n \to \infty,$$

where $X(t)$ is a markovian gaussian process with mean 0 and covariance

$$C(s,t) = \frac{s^2(2-t)^2}{4} \quad \text{when } s \le t.$$

The error term can be deduced for the various expansion in [Louchard, 1987]. It appears that the relative error in the density is $O(\frac{1}{\sqrt{n}})$ (non uniform in X). Thus, in this case, using the notations of Daniels' theorem, we have $z(t) = t(2-t)/2 - 1/2$, with maximum at $\bar{t} = 1$ and $z''(\bar{t}) = -1$. Since the covariance here is such a simple function, we can easily apply the theorem, and we find:

THEOREM 3.10 *If $Y_n(t)$ denotes the size of a random priority file history of length $2n$ at time $\lfloor nt \rfloor$, then we have:*

$$E(Max_t Y_n(t)) = \frac{n}{2} + \lambda n^{1/3} + O(n^{1/6}),$$

and more generally

$$Max_t Y_n(t) = \frac{n}{2} + \sqrt{n}\, M + O(n^{1/6})$$

is reached at time t^*, with M and t^* given by Daniels' theorem. We can prove that the error term in the weak convergence to $X(t)$ is negligible with the error in $MaxY$.

2.5.3 LIMITING PROFILES OF LIST STRUCTURES

In the previous section, we analyzed the distribution of the maximum of the priority queue . We can also get information on the limiting profile , i.e. on how much time is spent at each level k (k fixed, n going to infinity). Our investigations are based on the assumption that the size $Y(\lfloor nt \rfloor)$ of the list at time nt (with $0 \le t \le 2$) satisfies a weak convergence property:

$$\frac{Y(\lfloor nt \rfloor) - ny(t)}{\sqrt{n}} \Rightarrow X(t), \qquad 0 \le t \le 2$$

where y is a symmetric function around 1 and X is a gaussian process with mean 0 and known covariance C. In all applications to either classical or Knuth-type file histories, that assumption is true.

Let k, the level under study, be fixed, and $\bar{t}$ be the time such that $y(\bar{t}) = k/n$, with $\bar{t} < 1$.

If we consider the time u where Y first hits level k, the density of u is given by

$$\begin{aligned} g(u)du &= P(min\{u', Y(\lfloor nu' \rfloor) = k\} \in du) \\ &= P(min\{u', X(u') = \sqrt{n}(\frac{k}{n} - y(u'))\} \in du), \end{aligned}$$

and so

$$u - \bar{t} = O(\frac{1}{\sqrt{n}}).$$

Thus we examine the behaviour of Y near $\bar{t}$. Locally y can be approximated by a straight line, so that, $y'(\bar{t})$ denoting the slope of y at $\bar{t}$, we have

$$g(u)du \sim P(min\{u', X(u') = \sqrt{n}(\bar{t} - u')y'(\bar{t})\} \in du).$$

But it is known that locally X behaves like a Brownian Motion [Durbin, 1985]. Using classical results on the crossing time of a Brownian motion and a straight line, Cox and Miller [Cox and Miller, 1965] (p.221), and an asymptotic analysis, we find that the density of $\tau := \sqrt{n}(u - \bar{t})$ is just

$$\frac{y'(\bar{t})}{\sqrt{2\pi C(\bar{t}, \bar{t})}} e^{-(\tau y'(\bar{t}))^2 / 2C(\bar{t}, \bar{t})} d\tau,$$

where C is the covariance of X. Thus u is a gaussian variable centered at $\bar{t}$.

We now study the *total time spent* at level k. Let $p(u), q(u), r(u)$ be the probabilities that the next move on the list is an insertion, deletion and q respectively, if we start at $Y([nu])$ at time u. The random walk describing the evolution of the data structure is transient, so that between the time when the size of the structure first hits k, and the time when it last leaves level k, we may consider that $p(u), q(u)$ and $r(u)$ are locally constant and equal to the probabilities p, q and r of insertion, deletion and query at level k.

If the first step leaving level k is a deletion, we are sure of coming back to level k (since $\bar{t} < 1$). If it is an insertion, then the probability of ever returning to level k is q/p. If we look at the whole file history after hitting level k, the history spends l steps at level k, i steps inserting from k to $k+1$, d steps deleting from k to $k-1$, plus various other operations at other levels.

Thus, if $F(z, w, v)$ is the joint multidimensional generating function

$$F(z, w, v) = \sum_{i,d,l} P(l, i, d) z^l w^i v^d,$$

we have when $n \to \infty$:

$$F(z, w, v) = qzvF(z, w, v) + pw[\frac{q}{p}zF(z, w, v) + (1 - \frac{q}{p})] + rzF(z, w, v),$$

which leads to

$$F(z, w, v) = \frac{(p - q)w}{1 - qz(v + w) - rz} \tag{3.45}$$

All distributions we need can be extracted from Eq.(3.45). For instance, the time ℓ spent at level k is characterized by (let $w = v = 1$)

$$F_\ell(z) := \frac{p - q}{1 - (2q + r)z} \tag{3.46}$$

which shows that ℓ is a geometric R.V., with mean $E(\ell) = \frac{2q+r}{p-q}$. The *total* time spent at k has mean $\tilde{E}_\ell := 1 + E(\ell) = \frac{1}{p-q}$.

The number i of insertions steps *from* k is characterized by (let $z = v = 1$)

$$F_i(w) = \frac{(p-q)w}{p - qw} = \frac{(1 - q/p)w}{1 - wq/p}$$

This gives

$$P(i = \kappa) = \left(1 - \frac{q}{p}\right)\left(\frac{q}{p}\right)^{\kappa-1}, \qquad \text{(of course } i \geq 1) \qquad (3.47)$$

with mean

$$E(i) = \frac{1}{1 - q/p} \qquad (3.48)$$

Finally the number d of deletion steps *from* k is characterized by (let $z = w = 1$)

$$F_d(v) = \frac{p - q}{p - qv} = \frac{1 - q/p}{1 - vq/p} \qquad (3.49)$$

which shows that d is a geometric R.V. with mean

$$E(d) = \frac{q/p}{1 - q/p} \qquad (3.50)$$

2.5.4   EXAMPLE: CLASSICAL PRIORITY QUEUES

For priority queues, we have for the average size at t:

$$y(t) = \frac{1}{2}t(2 - t),$$

and the covariance of X is given by $C(s,t) = s^2(2-t)^2/4$. Thus $\bar{t} = 1 - \sqrt{1 - 2k/n}$, and $y'(\bar{t}) = \sqrt{1 - 2k/n}$. Adapting now the proof of Lemma 13 in [Louchard et al., 1992], we can prove that

$$p = 1 - \frac{\bar{t}}{2} + O(\frac{1}{n}), \quad q = \frac{\bar{t}}{2} + O(\frac{1}{n}), \quad r = 0.$$

Then we find that

$$E(l) = \frac{1 - \sqrt{1 - 2\frac{k}{n}}}{\sqrt{1 - 2\frac{k}{n}}} \text{ and } E(i) = E(d) = \frac{1}{\sqrt{1 - 2\frac{k}{n}}}.$$

3.   PROBABILISTIC ANALYSIS OF BINARY SEARCH TREES

Binary search trees are widely used to store (totally ordered) data. See H. Mahmoud's book [Mahmoud, 1992] for an excellent overview on this matter.

We have shown (cf. page 59) that the analysis of Quicksort can be reduced and performed on binary search trees built by successive insertions. In this section we focus on the probabilistic analysis of the height and the width of binary search trees.

3.1   HEIGHT OF BINARY SEARCH TREES

The probabilist analysis of the height of binary search trees has been done mostly by L. Devroye. [Devroye, 1986, Devroye, 1987] (see also [Mahmoud, 1992]). He proved that:

THEOREM 3.11 *Let h_n be the height of the binary search tree T_n,*

$$\frac{h_n}{\log n} \to^P c$$

where $c \simeq 4.311$ is a solution of the equation $x \log \frac{2e}{x} = 1$.

3.2   WIDTH OF BINARY SEARCH TREES

The following presentation is based on recent works of B. Chauvin, M. Drmota and J. Jabbour-Hattab (CDJ) [Chauvin et al., 1999]. Their paper is, so far, the best and most advanced analysis of binary search trees.

Remember that starting from a permutation of $\{1, 2, \dots, n\}$ we get a binary search tree with n (internal) nodes such that the keys of the left subtree of any given node x are smaller than the key of x and the keys of the right subtree are larger than the key of x. Usually it is assumed that any permutation of $\{1, 2, \dots, n\}$ is equally likely. CDJ use an alternative way of looking. They consider a Markov chain (cf. page 10) of trees which describes the evolution of a binary search tree. Instead of considering a permutation $\{1, 2, \dots, n\}$ they consider independent random variables $X_1, X_2, \dots$ which are uniformly distributed on $[0, 1]$. X_1 is associated to the root. If $X_2 < X_1$ then X_2 is associated to the left child of the root, and if $X_2 \geq X_1$ then X_2 is associated to the right child of the root and so on.

If we consider n random variables $X_1, X_2, \dots, X_n$ we get the same random model on binary trees with n internal nodes as the model induced by uniformly distributed random permutations on $\{1, 2, \dots, n\}$. Every parameter $Y = Y_T$ on binary trees T (i.e. height, width. etc.) induces a sequence $(Y(n))_{n\in\mathcal{N}}$ of random variables where $Y(n)$ denotes Y conditioned on the number n of internal nodes. Results on the height of binary search trees have been proven by Devroye in [Devroye, 1986, Devroye, 1987]. We present CDJ's approach [Chauvin et al., 1999] for proving that:

THEOREM 3.12 *The width V_n (the maximal number of internal nodes at the same level) satisfies*

$$V_n \sim \frac{n}{\sqrt{4\pi log n}} \ as \ n \to \infty.$$

Proof: We present a sketch of CDJ's proof. It appears that martingales (page 19) play here again a central role. U_k stands for the number of external nodes at level k, V_k for the number of internal nodes at level k and $Z_k = U_k + V_k$ the total number of nodes at this level. It is easy to see that:

- $Z_{k+1} = 2V_k$.

- $Z_{k+1} - Z_k = V_k - U_k$.

- $Z_k = \sum_{j\geq k} 2^{k-j} U_j$

The main tool for their proof is the power series

$$W_n(z) = \sum_{j\geq 0} U_k(n) z^k$$

We summarize below preliminary results which are corner stones of the proof.

LEMMA 10 $\frac{W_n(z)}{EW_n(z)}$ *is a martingale with respect to the natural filtration $\mathcal{F}_n$ associated to the sequence of trees T_n,*

$$EW_n(z) = \prod_{j=0}^{n-1} \frac{j + 2z}{j + 1}$$

For any compact set C in the complex plane C we have:

$$EW_n(z) = \frac{n^{2z-1}}{\Gamma(2z)} + O(n^{2Re(z)-2})$$

Then the idea of the proof is to use a.s. expansion of $W_n(z)$. See the details in [Chauvin et al., 1999].

4.   PROBABILISTIC ANALYSIS OF GENETIC ALGORITHMS

Genetic algorithms are search algorithms based on the genetic mechanisms which guide natural evolution: mutation, crossover and selection. They have been developed by J.H. Holland [Holland, 1975] in the seventies. Today, genetic algorithms are widely used to handle a large number of optimisation problems ranging from the classical traveling salesman problem to the design of network architecture. D. Goldberg's book [Goldberg, 1989] is an excellent introduction to this area and presents many applications of genetic algorithms in search, optimization and machine learning. Experimental simulations are often used for handling concrete problems. R. Cerf [Cerf, 1994] has developed a probabilistic framework (based on large deviations and Freidlin-Wentzell's theory) for the analysis of genetic algorithms. We present below a simple, classical genetic algorithm [Goldberg, 1989, Cerf, 1994]

4.1   THE CLASSICAL GENETIC ALGORITHM

The algorithm under consideration is an inductive, stochastic algorithm which operates on sets of points and is designed with the help of three operators: mutation, crossover (i.e. crossing-over) and selection.

The set of points (called population) is taken from the space $\{0,1\}^N$. N is, in general, too large for an exhaustive search. Let $m \in \mathcal{N}$ be the (fixed) size of the population and X_n the population at time n:

$$X_n = (X_n^1, \dots, X_n^m)$$

and the components (i.e. chromosomes) are words of length N on the alphabet $\{0,1\}$.

The transition from X_n to X_{n+1} decomposes into three steps:

$$X_n \xrightarrow{mutation} Y_n \xrightarrow{crossover} Z_n \xrightarrow{selection} X_{n+1}$$

We now describe each step.

Mutation: $X_n \longrightarrow Y_n$.

A parameter p between 0 and 1 is initially chosen: p is the probability of mutation. For each letter of each chromosome $XN^1, \dots, X_n^m$ a p parameter Bernoulli trial is done and, depending of the result, the associated letter is changed or not (i.e. with probability p, the letter is replaced by the other one $(0 \rightarrow 1, 0 \rightarrow 1)$ and is remains unchanged with probability $1 - p$). In general p is a small number. The new resulting population is denoted Y_n.

Crossover: $Y_n \longrightarrow Z_n$.

Here also a parameter q beteween 0 and 1 is fixed. q is the crossover probability. For constructing the population Z_n, we form $\frac{m}{2}$ pairs with the items of Y_n (for example by choosing the items randomly or choosing direct neigbors). Now we perform a Bernoulli trivial with parameter q in order to decide if the crossover will be realized. If yes, a cutting position is chosen randomly between 1 and $N - 1$ and the end parts of two chromosomes are permuted. We get a new pair of chromosomes which

$$
\begin{array}{c}
010 \underline{\quad} 00110 | 011100 \underline{\quad\quad} 011 \\
111 \underline{\quad} 01101 | 111101 \underline{\quad\quad} 101
\end{array}
\longrightarrow
\begin{array}{c}
0\,10 \underline{\quad} 00110 | 111\,101 \underline{\quad} 101 \\
1\,11 \underline{\quad} 01101 | 011100 \underline{\quad} 011
\end{array}
$$

Figure 3.4. Crossover

is part of the population Z_n. In general q is close to 1.
Selection: $Z_n \longrightarrow X_{n+1}$.
The m chromosomes of the population X_{n+1} are selected independently with the help of a probability distribution function on the population Z_n (i.e. a selection function). Let f be the map $\{0,1\}^N \to \mathcal{R}_+^*$ for which we want to dertermine the global maxima. The most common selection function is given by

$$
\forall h \in \{1,\ldots,m\}, \quad P(Z_n^h) = \frac{f(Z_n^h)}{f(Z_n^1) + \ldots + f(Z_n^m)}
$$

EXAMPLE 14 *Consider the function* $f(x) = \sqrt{x - x^2}$, $x \in [0,1]$. *We show how to determine the trivial* 0.5 *maxima of* f. *The representation of the solutions depends on the precision we want. For example, if we want a* 0.001 *precision, we divide the interval* $[a,b] = [0,1]$ *into* $\frac{b-a}{0.001}$ *intervals of size* 0.001. *Since* $2^9 < 1000 < 2^{10}$, 10 *bits are required for the representation of the variables. We get therefore:*

$$
a = 0 = 0000000000, \quad b = 1111111111
$$

and $x \in [a,b]$ *such that*

$$
a + k\frac{b-a}{2^{10}} < x < a + (k+1)\frac{b-a}{2^{10}}
$$

where k *belongs to* $[0, 2^{10}]$, *is represented by the binary decomposition of* k. *Example, for* $x = 0.681$, *we get* $k = 697$. *The binary representation of* x *is* $(697)_2 = 1010111001$ *and we write* $0.691 = 1010111001$
Choose a pair:

$$
0.681 = 101|0111001, \quad f(0.681) = 0.466
$$

$$0.885 = 111|0001010, \quad f(0.885) = 0.319$$

$$101|0001010 = 0.635, \quad f(0.635) = 0.481$$

$$111|0111001 = 0.931, \quad f(0.931) = 0.253$$

The crossover was benefic since for 0.635 we get a value closer to 0.5 (and the very bad value 0.253!) Step after step, do we get closer to the maxima 0.5?. Experimentations seem to be in favor of a positive answer. What about the choice of the parameters and the crossover? Do they affect the rate of convergence? These are the fundamental questions investigated by R. Cerf [Cerf, 1994].

4.2   PROBABILISTIC ANALYSIS: SOME HINTS

4.2.1   A MARKOV CHAIN MODEL

Remember that X_n represents the population at time n. The series $(X_n), n \in \mathcal{N}$ is a Markov chain (4.) with state space $(\{0,1\}^N)^m$. The distribution law of X_n is uniquely determined by the distribution of X_0 (in general uniform on $(\{0,1\}^N)^m$ and the mechanism described in the previous section. This mechanism is very complicated but has the following fundamental properties

- It is homogeneous (i.e. does not depend on n)

- It is irreducible: Given two populations x and y, the probability starting from x to get y in a finite number of transitions is strictly positive

$$\forall x, y \in (\{0,1\}^N)^m, \; \exists r \in \mathcal{N}, \; P(X_{n+r} = y | X_n = x) > 0.$$

This means that the full space of the populations will be explored.

- It is aperiodic

$$\forall x \in (\{0,1\}^N)^m, \; GCD\{r : P(X_{n+r} = x | X_n = x) > 0\} = 1$$

Then we can apply the following result concerning Markov chains (4.).

THEOREM 3.13 *An homogeneous, irreducible, aperiodic Markov chain with finite state space is ergodic and has a unique stationary (i.e. invariant) probability measure.*

4.2.2   ASYMPTOTIC CONVERGENCE

As mentioned previously, R. Cerf [Cerf, 1994] has provided the most rigorous mathematical analysis of genetic algorithms. We present below his model and his analysis of the simple genetic algorithm.

Let E be a finite set and f a positive function defined on E. We wish to find the set f^* of the global maxima of f (called the fitness function). Let m be a population size. For an $m-uple$ $x = (x_1, \ldots, x_m)$ of E^m he uses the following notations:

$$[x] = \{x_k : 1 \leq k \leq m\},$$

$$\hat{x} = \{x_k : 1 \leq k \leq m, f(x_k) = max_{1 \leq h \leq m} f(x_h)\}$$

$$x(i) = card\{k : 1 \leq k \leq m, \ x_k = i\}$$

The set A is the set of all m-uples whose m components are equal (such components are called uniform populations). S denotes the set of populations whose individuals have the same fitness:

$$S = \{x \in E^m : \ f(x_1) = \ldots = f(x_m)\}$$

Let (X_n^∞) be the Markov chain with state space E^m having for transition probabilities

$$
\begin{aligned}
P(X_{n+1}^\infty = z | X_n^\infty = y) &= \frac{1}{(card\hat{y})^m} \prod_{k=1}^{m} 1_{\hat{y}}(z_k) y(z_k) \\
&= \frac{1}{(card\hat{y})^m} \prod_{i \in [z]} 1_{\hat{y}}(i) y(i)^{z(i)}
\end{aligned}
$$

That is, the individuals of the population X_{n+1}^∞ are chosen randomly (under the uniform distribution) and independently among the elements of $\hat{X}_n^\infty$ which are the best individuals of X_n^∞ according to the fitness function f. R. Cerf builds a sequence of Markov chains (X_n^l) by perturbing randomly (X_n^∞). The integer parameter l describes the intensity of the perturbations. The chain (X_n^l) is obtained through the three chain buildings

$$X_n^l \xrightarrow{mutation} U_n^l \xrightarrow{crossover} V_n^l \xrightarrow{selection} X_{n+1}^l$$

The mutation operator $X_n^l \longrightarrow U_n^l$ is modelled by random independent perturbations of the individuals of the population X_n^l. The transition probabilities from X_n^l to U_n^l are then given by

$$P(U_n^l = u | X_n^l = x) = p_l(x_1, u_1) \ldots p_l(x_m, u_m)$$

where p_l is a markovian kernel on the space E satisfying
Hypothesis H_p. There exists a function α defined on $E \times E$ with values in $\mathcal{R}^+$ which is an irreducible kernel, i.e.

$$\forall i, j \in E; \ \exists i_1, \ldots, i_r; \ i_0 = i, i_r = j, \quad \prod_{0 \leq k \leq r-1} \alpha(i_k, i_{k+1}) > 0$$

and a positive real number a such that p_l admits the development

$$\forall i, j \in E, \ \forall s \leq a$$
$$p_l(i,j) = \begin{cases} \alpha(i,j)l^{-a} + 0(l^{-s}) & \text{if } i \neq j \\ 1 - \alpha(i,j)l^{-a} + 0(l^{-s}) & \text{if } i = j \end{cases}$$

The crossover $U_n^l \longrightarrow V_n^l$ is modelled by random independent perturbations of the pairs formed by consecutive individuals of the population $(V_n^l$. The transition probabilities from U_n^l to V_n^l are

$$P(V_n^l = v | U_n^l = u) = \delta_m(u_m, v_m) \prod_{1 \leq k \leq m/2} q_l((u_{2k-1}, u_{2k}), (v_{2k-1}, v_{2k}))$$

where $\delta_m(i,j) = \delta(i,j)$ if m is odd (the last individual remains unchanged), $\delta_m(i,j) = 1$ if m is even, and q_l is a markovian kernel on the space $E \times E$ satisfying
Hypothesis H_q . There exists a function β defined on $(E \times E) \times (E \times E)$ with values in $\mathcal{R}^+$ and a positive number b such that q_l admits the development
$\forall i_1, j_1 \in E, \ \forall i_2, j_2 \in E, \ \forall s \leq b,$

$$q_l((i_1, j_1), (i_2, j_2)) = \begin{cases} \beta((i_1,j_1),(i_2,j_2))l^{-b} + 0(l^{-s}) \\ \quad \text{if } (i_1,j_1) \neq (i_2,j_2) \\ 1 - \beta((i_1,j_1),(i_2,j_2))l^{-b} + 0(l^{-s}) \\ \quad \text{if } (i_1,j_1) = (i_2,j_2) \end{cases}$$

For the selection operator $V_n^l \longrightarrow X_{n+1}^l$, the transition probabilities are

$$P(X_{n+1}^l = x | V_n^l = v) = \prod_{i \in [x]} \left(\frac{v(i) exp(cf(i) \ln l)}{\sum_{k=1}^m exp(cf(v_k) \ln l)} \right)^{x(i)}$$

where $c \in \mathcal{R}_*^+$ is a scaling parameter.

As l grows toward infinity the perturbations become smaller and smaller and with overwhelming probability, the chain (X_n^l) behaves as (X_n^∞) would do: it is attracted by the set S. Using large deviations and more precisely Freidlin-Wentzell theory [Freidlin and Wentzell, 1984], R. Cerf [Cerf, 1994] carries out a precise study of the asymptotic dynamics of the chain when the perturbations disappear. Let T_n be the time of the n-th visit to the set S and put $Z_n^l = X_{T_n}^l$. The chain $(Z_n^l)_{n \in \mathcal{N}}$ is the Markov chain induced by the chain (X_n^l) on the set S. The study of the chain (Z_n^l) of successive visits to the set of attractors S is equivalent to the study of the whole process (X_n^l) (see [Freidlin and Wentzell, 1984], chapter 6). It follows from the assumptions that we have

$$P(Z_{n+1}^l = z | Z_n^l = y) \sim \tilde{C}(y, z) l^{-\tilde{V}(y,z)} \quad as \quad l \to \infty$$

where $\tilde{C}$ and $\tilde{V}$ are two non-negative kernels on $S \times S$. These estimates yield estimates of the invariant measure of the chain (Z_{n+1}^l) through the representation formula involving Freidlin-Wentzell graphs (see [Freidlin and Wentzell, 1984], chaper 6). The crucial quantity is the virtual energy W, defined for x in S by

$$W(x) = min\{ \sum_{(y \to z) \in g} \tilde{V}(y, z) : g \in G\{x\}\}$$

where $G\{x\}$ is the set of all x-graphs over S. Two situations are then considered. In the homogeneous case, R. Cerf studies the symptotic behavior of the stationary measure of (X_n^l). As l goes to infinity, this stationary measure concentrates on the set

$$W^* = \{x \in S : W(x) = min_{y \in S} W(y)\}\}$$

He shows that $W^* \subset f^*$ for a sufficiently large population size so that

$$\forall x \in E^m, lim_{l \to \infty} lim_{l \to \infty} P([X_n^l] \in f^* | X_0^l = x) = 1$$

In the inhomogeneous case, where l is an increasing function of n, he derives several conditions on the population size and on the rate of increasing of the sequence $l(n)$, some necessary and some sufficient, to ensure the settling in f^* in finite time, i.e.

$$\forall x \in E^m, P(\exists N, \forall n \geq N, [X_{T_n}] \subset f^* | X_0 = x) = 1$$

It remains to specialize R. Cerf's general model in order to apply his results to the classical simple genetic algorithm under consideration. Take $E = \{0, 1\}^N$ for some integer N. A point i of E is a word of length N over the alphabet $\{0, 1\}$ and has the form $i = i_1 \ldots i_N$ where $i_k \in \{0, 1\}$.

The Hamming distance $H(i, j)$ between two points i, j of E is the number of letters where i and j differ:

$$H(i, j) = card\{k : 1 \leq k \leq m, i_k \neq j_k\}$$

The mutation kernel p_l and its associated kernel α are

$$p_l(i,j) = \begin{cases} 0 & \text{if } H(i,j) > 1 \\ l^{-a} & \text{if } H(i,j) = 1 \\ 1 - Nl^{-a} & \text{if } H(i,j) = 0 \end{cases}$$

$$\alpha(i,j) = \begin{cases} 0 & \text{if } H(i,j) > 1 \\ 1 & \text{if } H(i,j) = 1 \\ N & \text{if } H(i,j) = 0 \end{cases}$$

Two arbitrary points of E are connected through the kernel α in at most N transitions so that the irreducibility condition is satisfied.

In order to build the crossover operator, R. Cerf befines $N-1$ cutting operators $(T_k)_{1 \leq k < N}$. T_k maps $E \times E$ onto $E \times E$ and for i, j in E he puts $T_k(i,j) = (i',j')$ where

$$i' = i_1 \ldots i_k j_{k+1} \ldots j_N, \quad j' = j_1 \ldots j_k i_{k+1} \ldots i_N$$

The kernel β is given by:

$$\beta((i,j),(i',j')) = card\{k : 1 \leq k \leq N-1, T_k(i,j) = (i',j')\}$$

and the kernel q_l is defined by:

$$q_l((i,j),(i',j')) = \beta((i,j),(i',j'))l^{-b} \quad if \ (i,j) \neq (i',j'),$$

$$q_l((i,j),(i,j)) = 1 - \sum_{(i',j') \neq (i,j)} \beta((i,j),(i',j'))l^{-b}$$

Both hypotheses H_p and H_q are fulfilled.

Let

$$\Delta_1 = max_{i \in (E-f^*)} min\{max_{0 \leq k < r}(f(i) - f(i^k)) : i^0 = i, r \leq N, f(i^r) > f(i),$$
$$\prod_{0 \leq k < r} H(i^k, i^{k+1}) > 0\}$$

$$\Delta_2 = max_{i,j \in f^*} min\{max_{0 \leq k < r}(f(i) - f(i^k)) : i^0 = i, r \leq N, i^r = j,$$
$$\prod_{0 \leq k < r} H(i^k, i^{k+1}) > 0\}$$

and let

$$\Delta_3 = \max(\Delta_1, \Delta_2), \quad \delta = \min\{|f(i) - f(j)| : i,j \in E, f(i) \neq f(j)\}$$
$$(3.51)$$

Then R. Cerf [Cerf, 1994] proves the following convergence result for the homogeneous simple genetic algorithm:

THEOREM 3.14 *If*

$$m > \frac{aN + c(N-1)\Delta_3}{min(a, b/2, c\delta)}$$

then

$$\forall x \in E^m, \ lim_{l\to\infty}lim_{n\to\infty}P([X_n^l] \subset f^* | X_0^l = x) = 1.$$

The next result involves the convergence exponent of an increasing sequence $l(n)$ i.e. the unique real number λ such that the series $\sum_{n\geq 0} l(n)^{-\theta}$ converges for $\theta > \lambda$ and diverges for $\theta < \lambda$.

For the inhomogeneous simple genetic algorithm he gets [Cerf, 1994]:

THEOREM 3.15 *1) For the chain (X_{T_n}) to be trapped in f^* after a finite number of transitions, i.e. to have*

$$\forall x \in E^m, \ P(\exists N, \forall n \geq N, [X_{T_n}] \in f^* | X_0 = x) = 1$$

the convergence exponent of the sequence $l(n)$ must be a positive real number; that is there must exist two positive real numbers θ_1 and θ_2 such that

$$\sum_{n\geq 0} l(n)^{-\theta_1} = \infty \ \ and \ \ \sum_{n\geq 0} l(n)^{-\theta_2} < \infty$$

2) If the convergence exponent λ of the sequence $l(n)$ and the population size m satisfy the inequalities

$$aN + c(N-1)\Delta_3 < \lambda < min(a, b/2, c\delta)m$$

then, with probability one, the chain (X_{T_n}) is trapped in f^* after a finite number of transitions, i.e.

$$\forall x \in E^m, \ P(\exists N, \forall n \geq N, [X_{T_n}] \in f^* | X_0 = x) = 1$$

3) Suppose that there exists a real number t strictly greater than one such that for all $r \in N$, the sequences $l([tn] + r)$ and $l(n)$ are logarithmically equivalent. If the convergence exponent λ of the sequence $l(n)$ and the population size m satisfy the inequalities

$$aN + c(N-1)\Delta_3 < min(a, b/2, c\delta)m \leq \lambda$$

then

$$\forall x \in E^m, \ lim_{n\to\infty}P([X_n] \in f^* | X_0 = x) = 1.$$

5.   A MARKOVIAN CONCURRENCY MEASURE

In this section we present another application of Markov chains. More precisely, we modelize concurrency between several processors in terms of automata and Markov chains. Then we define a concurrency measure which reflects faithfully the behaviour of the processes and is easy to compute with a symbolic manipulator. The material of this section comes from [Geniet et al., 1996].

5.1   GENERALITIES

The notion of concurrency plays a central role in parallel processing and is studied extensively mostly from algebraic and semantic points of view. A combinatorial approach has been given by J. Françon [Françon, 1986]: he shows how to count explicitly the number of correct (i.e. without deadlock) behaviours of parallel systems under the mutual exclusion policy. His concurrency measure is the inverse of the radius of convergence of some generating function. J. Beauquier, B. Bérard and L. Thimonier [Beauquier et al., 1987] consider also all behaviours of the concurrent systems and their measure, based on Arnold-Nivat's model [Arnold and Nivat, 1982], is the average waiting time of the processes. The measure presented in this section uses also Arnold-Nivat's model . The behaviour of concurrent systems is modellizeded by absorbing Markov chains whose transition matrices contain all information necessary for easy computation of the concurrency measure. Then we apply the properties of the fundamental matrix (cf. page 12). The computing of the concurrency measure reduces to simple linear algebra computations.

5.2   GENERALITIES ON LANGUAGES AND AUTOMATA

Let Σ^* be the free monoid over the alphabet Σ and w a word of $L \in \Sigma^*$. $|w|$ stands for the length of w, $|w|_x$ for the number of occurrences of x in w (x may be a letter or a subword) and $w^{(i)}$ is the $i - th$ letter of w. A language is regular if it is recognized by an automaton (see definition below) with a finite number of states. Let $(L_i)_{i \in [1,n]}$ be a family of regular languages defined over the alphabets $(\Sigma_i)_{i \in [1,n]}$. Let $r_i : \prod_{j=1}^n \Sigma_j \to \Sigma_i$ be the i-th projection. We extend r_i to an homomorphism

$$r_i : (\prod_{j=1}^n \Sigma_j)^* \to (\Sigma_i)^*$$

$$\Omega_{j=1}^{n} L_j = \{w \in (\prod_{j=1}^{n} \Sigma_j)^* | \forall i \in [1,n], r_i(w) \in L_i\}$$

is called homogeneous product.

It follows that $\forall i, |r_i(w)| = |w|$. Note that the homogeneous product of regular languages is always regular.

Let $(\Sigma_i)_{i \in [1,n]}$ be a family of alphabets and Σ be the product of these alphabets (a letter of Σ is a vector whose $i - th$ component is a letter of Σ_i). We call projector on the $i - th$ component the morphism that associates with an element w of Σ its $i - th$ component w_i. Let us remember that:

DEFINITION 3.3 *A finite automaton is a 5-uple* $(\Sigma, Q, S, F, \delta)$ *where* Σ *is an alphabet,* $Q \subset N$ *the set of states,* $S \subset Q$ *and* $F \subset Q$ *the sets of initial and terminal states and* $\delta \subset Q \times \Sigma \times Q$ *the set of transitions.*

REMARK 13 *The product of n automata is defined as follows: the set of states is the product of the sets of states, the set of transitions is the product of the sets of transitions, the terminal state is the product of the terminal states.*
The product accepts the homogeneous product of the languages L_{A_i} *[Arnold, 1994]. Notation.*
For each $t = (i, w, j)$ *of* δ, *we denote by* t_- *the state* i, t_+ *the state* j *and* w_t *the label* w. *For each* i *of* Q, *we denote* i_- *the set of transitions* t *of* δ *such that* $t = (x, w, i)$ *and* i_+ *the set of transitions* t *of* δ *such that* $t = (i, w, x)$.

DEFINITION 3.4 *A finite probabilistic automaton is a 5-uple* $(\Sigma, Q, S, F, \delta)$, *where* Σ *is an alphabet,* $Q \subset N$ *the set of states,* $S \in Q$ *the set of initial states and* $F \in Q \times [0,1]$ *the set of terminal states and the corresponding absorption probabilities,* $\delta \in Q \times \Sigma \times [0,1] \times Q$ *the set of transitions and, if* $p(t)$ *(resp.* $p(i)$*) is the probability associated with the transition* t *(resp. the terminal state* i*), we get* $i \notin F \Rightarrow \sum_{t=i+} p(t) = 1$, *and* $i \in F \Rightarrow \sum_{t=i+} p(t) = 1 - p(i)$.

REMARK 14 i *is an absorbing state if* $p_{i,i} = 1$, *where* $p_{i,i}$ *is the probability to remain in the state* i.

5.3 REPRESENTATION OF CONCURRENT SYSTEMS

First we show how to modelize concurrent processes.

5.3.1   ARNOLD-NIVAT'S MODEL

In this model, a system of processes is represented by a synchronized product, i.e. an homogeneous product, where the transitions labelled by some forbidden vectors have been removed (see [Arnold and Nivat, 1982, Arnold, 1994]).

5.3.2   CONSTRUCTION OF A SYNCHRONIZED SYSTEM

The fact that a process is blocked is modeled by the symbol #, and L_{A_i} is now replaced by $L_{A_i}\amalg\{\#\}^*$, where II stands for the shuffle product. The concurrency measure will correspond to the average number of #'s over all words accepted by the automaton.

5.3.3   PROBABILISTIC AUTOMATON ASSOCIATED WITH ARNOLD-NIVAT'S MODEL

The basic idea is to transform the previous automaton into a probabilistic one. The transitions are labelled with probabilities whose values are obtained through experiments on the program. The assumptions are these of the synchronized product.

Computation of the transition probabilities.   The probabilities associated with the transitions of the product automaton are computed from the probabilities of the transitions of the automata associated with the sequential processes.In the folowing, we consider a transition t of the product automaton to be issued from a state s of the product automaton. t is the vector $(t_i)_{i\in[1,n]}$, each t_i being a transition of the automaton A_i. We denote $S_t = \{i \in [1,n] | w_t^{(i)} = \#\}$, and

$$\prod(t) = \prod_{i\in[1,n]-S_t} p(t_i)$$

Let $\phi(t) = \sum_{t\in i_+} \prod(t)$ be the output probabilistic flow of state i. The probability of each t issued from i is the output flow the value $\prod(t)$ represents in the set i_+:

$$p(t) = \frac{\prod(t)}{\Phi}$$

The product of a family of probabilistic automata is a probabilistic automaton: its algebraic structure is the homogeneous product of the algebraic structures of the component probabilistic automata; the probabilities associated to its transitions fulfil the conditions of Definition

3.4:

$$p(i) + \sum_{t\in i_+} p(t) = \sum_{t\in i_+} \frac{\prod(t)}{\Phi}$$

$$= \frac{1}{\Phi} \sum_{t\in i_+} \prod(t)$$

$$= \frac{\Phi}{\Phi} = 1$$

(one can see that this property holds for terminal states too).
The effective determination of the probabilities is very simple for imperative statements (i.e. $p(t) = 1$), but must be detailed for test or loop statements as well as for absorption.

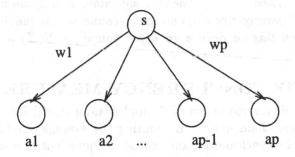

Figure 3.5. Flowchart automaton 1

Tests and absorption. In the flowchart automaton A, a test statement $If\ldots Then\ldots Else$ or $Case\ldots Of\ldots End$ generates a state s with many output transitions $(t_i)_{i\in[1,p]}$: $|s_+| > 1$ (see Figure 3.5). We determine the corresponding probabilities in the following way: the program is executed N times, and we count the number of times the transition t_i is performed. Then we take:

$$p(t_i) = \frac{N_i}{\sum_{i=1}^{p} N_j}$$

Absorption is treated in the same way: we count the number of times a terminal state is reached.

Loops. A loop can be represented by the diagram.
Statistics on the average number of looping steps performed by real programs lead us to consider that the random variable X (equal to the

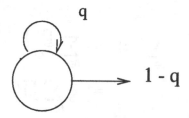

Figure 3.6.   Flowchart automaton 2

number of steps required to leave the loop) has a geometric (Pascal) distribution: $p(X = n) = q^{n-1}(1 - q)$ (with $q \in [0,1]$), hence $E(X) = \frac{1}{1-q}$.

REMARK 15 *When the implemented algorithm is well-kown, the study of its speed of convergence gives an upper-bound for the maximal number of steps. From this we deduce an upper bound for $E(X)$ and therefore for q.*

5.4   THE CONCURRENCY MEASURE

The probabilistic extension of Arnold-Nivat's model leads to an absorbing Markov chain model. Evaluating the average length of a word accepted by the synchronized automaton is equivalent to evaluating the length of the paths before absorption in the corresponding Markov chain.

Let $\bar{n}$ (resp. $\bar{a}$) be the average number of steps (resp. #'s) until absorption, and p the number of processes.

DEFINITION 3.5 *The concurrency measure of the system is the ratio:*

$$\frac{\bar{a}}{p\bar{n}} \tag{3.52}$$

5.5   COMPUTATION OF THE CONCURRENCY MEASURE

We use the results of Section 1 about the fundamental matrix of absorbing Markov chains. As stated previously, the computation of the concurrency measure Eq.(3.52), involves only classical operation on matrices and can be easily done with a symbolic manipulator.

5.6   EXAMPLE: MUTUAL EXCLUSION

We illustrate with the mutual-exclusion problem. The corresponding language is $L = (r^*d\#^*c + s)^*$, where the actions are r (not critical section), d (asking for the resource), # (waiting when the resource is used by another process), c (critical section) and s (leaving the critical section).

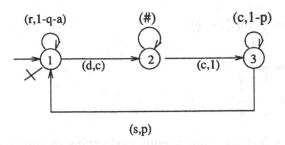

Figure 3.7. Flowchart automaton 3

For simplicity, we consider that all these processes begin and terminate their execution simultaneously. L is accepted by the automaton given in Figure 3.7.

Construction of the synchronized automaton

Let:

- $\bar{\gamma}$ be the average time of each owning of the critical section .

$$\bar{\gamma} = 1 + E(\text{Number of transitions from state 3 to 3} \mid$$
$$\text{The chain is in state 3})$$
$$= 1 + p \sum_{i=0}^{\infty} i(1-p)^i$$
$$= \frac{1}{p}$$

- $\bar{\rho}$ be the average time spent in the critical section.

$$\bar{\rho} = 1 + E(\text{Number of transitions from state 1 to 1} \mid$$
$$\text{The chain is in state 1})$$
$$= \sum_{i=0}^{\infty} i(1-q-a)^i(q+a)$$
$$= \frac{1}{q+a} - 1$$

- $\bar{\alpha}$ be the average number of requests for the critical section

$$\bar{\alpha} = 1 + E(\text{Number of transitions to state 2} \mid$$
The chain is in state 1)

$$= \sum_{i=0}^{\infty} i q^i (1 - q)$$

$$= \frac{1}{1 - q} - 1$$

$$= \frac{q}{1 - q}$$

Synchronisation forbids simultaneous performing of the critical section by two different processes. Therefore, all transitions whose label contains more than one occurrence of the letter c must be deleted.

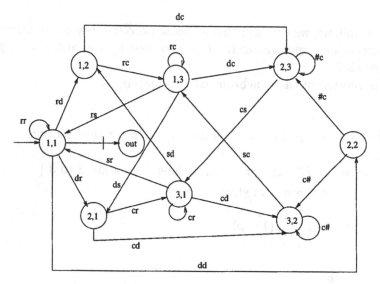

Figure 3.8.   Flowchart automaton 4

The synchronized automaton of two processes is drawn on Figure 3.8. The transitions labelled $(c, \#)$ and $(\#, c)$ from state $(2, 2)$ respectively to states $(3, 2)$ and $(2, 3)$ have the same probability: this is due to the fact that there is no priority for the processes.

The computation of the concurrency measure for different values of γ, ρ and α leads to different values for the efficiency of the concurrent system. Let $\bar{n}$ be the average waiting time before absorption. The table below shows several experimentations:

The performances are less efficient when processes stay a long time in the critical section. Therefore we conclude that the system is more efficient when there are many short jobs using the resource than when there are few long jobs.

6. PROBABILISTIC ANALYSIS OF BIN PACKING PROBLEMS

In this section we present Rhee-Talagrand's nice result on the bin packing problems [Rhee and Talagrand, 1987]. By the way [Rhee and Talagrand, 1987] contains also a result for the Traveling Salesman Problem. These problems can be modelized and analysed in terms of martingales (cf. page 19). The bin packing problem can be stated as follows: Find the minimum number of unit size bins needed to pack a given collection of items with sizes in $[0, 1]$.

Let n be the number of items to pack. The sizes if items are $X_1, \ldots, X_n$ independent random variables, identically distributed over $[0, 1]$. X_k will denote item name and item size.

Assume that the packing procedure has the following property (called H):

If the procedure packs $X_1, \ldots, X_n$ in k bins, the number of bins needed to pack $X_1, X_l, Y, X_{l+1}, \ldots, X_n$ for any l, and Y in $[0, 1]$ is at least k and at most $k + 1$.

THEOREM 3.16 *Let B be the number of bins needed to pack $X_1, \ldots, X_n$. If condition (H) holds, then for $t \geq 0$:*

$$P(|B - E(B)| > t) \leq 2e^{-\frac{t^2}{2n}}.$$

Proof: We give a sketch of Rhee-Talagrand's proof (see [Rhee and Talagrand, 1987] for more details).

For $0 \leq i \leq n$, let Σ_i be the σ-field generated by X_l, $l \leq i$. Then B is Σ_n measurable.

Fix i, $1 \leq i \leq n$. Denote by B^i the number of bins needed to pack $X_1, \ldots, X_{i-1}, X_{i+1}, \ldots, X_n$.

By property (H):

$$B^i \leq B \leq B^{i+1}$$

Then:

$$E(B^i|\Sigma_i) \leq E(B|\Sigma_i) \leq E(B^i|\Sigma_i) + 1$$

and

$$E(B^i|\Sigma_{i-1}) \leq E(B|\Sigma_{i-1}) \leq E(B^i|\Sigma_{i-1}) + 1$$

Howewer:

$$E(B^i|\Sigma_i) = E(B^i|\Sigma_{i-1})$$

Thus

$$|E(B|\Sigma_i) - E(B|\Sigma_{i-1})| \leq 1$$

Let $d_i = E(B|\Sigma_i) - E(B|\Sigma_{i-1})$. Then $E(d_i|\Sigma_{i-1}) = 0$ for each i. (d_i) is called a martingale difference sequence and we can apply Azuma's lemma:

LEMMA 11 *Let* (d_i), $1 \leq i \leq n$ *be a martingale sequence. Then for each* $t > 0$:

$$P(|\sum i \leq nd_i| > t) \leq e^{-\frac{t^2}{2\sum i \leq n\|d_i\|_\infty^2}}.$$

Now we apply the above lemma to $B - E(B) = \sum_{1 \leq i \leq n} d_i$ and remark that $\sum_{i \leq n} \|d_i\|_\infty^2 \leq n$ in order to end the proof.

As stated in [Rhee and Talagrand, 1987] extension to the infinite case is possible:
Assume that if $X_1, \ldots, X_n$ are packed in k bins and $Y_1, \ldots, Y_m$ are packed in l bins then $X_1, \ldots, X_n, Y_1, \ldots, Y_m$ are packed in at most $k+l$ bins. Let B_n be the number of bins needed to pack the first n items of an infinite independent, indentically distributed sequence (X_i).
Then Kingman's theory of subadditive processes tells us that [Ito and McKean, 1974, Stroock, 1984]:

$$lim_{n\to\infty}\frac{B_n}{n} = c \ a.s.$$

where c is a constant. And we get (see [Rhee and Talagrand, 1987]):

THEOREM 3.17 *For each* $\epsilon > 0$ *and* n *large enough:*

$$P(|\frac{B_n}{n} - c| > \epsilon) \leq 2e^{-\frac{n\epsilon^2}{4}}.$$

7.   PROBABILISTIC ANALYSIS OF SOME DISTRIBUTED ALGORITHMS

7.1   TWO STACKS PROBLEM

We consider the evolution of two stacks inside a shared, contiguous memory area of a fixed size m. The shared storage allocation algorithm studied in this section lets the stacks grow from the two ends of the memory until the cumulative size of the stacks exhausts the available storage. This algorithm is to be compared to another strategy, namely allocating separate zones of size $m/2$ to each stack. This problem of Knuth [Knuth, 1973] has been investigated by Yao [Yao, 1981], Flajolet [Flajolet, 1986], Louchard and Schott [Louchard and Schott, 1991] and Maier [Maier, 1991b]. Below we present the probabilistic analysis of Louchard and Schott [Louchard and Schott, 1991] then we focus on Maier's large deviations approach [Maier, 1991b] as we did in Section 2.4.2 for the analysis of dynamic data structures. The natural formula-

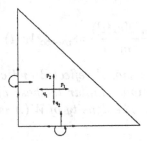

Figure 3.9.   Evolution of two stacks

tion of the two stacks problem is in terms of random walks in a triangle in a two-dimensional lattice space: a state is pair formed by the size of both stacks. The random walk $Y_m(.)$ has two reflecting barriers (R) along the axes ans one absorbing barrier (A) parallel to the second diagonal (see Figure 3.9). The distribution of steps (steps of unit length) ΔY is given by:

$$P(\Delta Y = (1,0)) = p_1, \ P(\Delta Y = (-1,0)) = q_1, \ P(\Delta Y = (0,1)) = p_2,$$
$$P(\Delta Y = (0,-1)) = q_2$$

with the boundary conditions of Figure 3.9. We assume that the transition probabilities are constant and that $p_1 + q_1 + p_2 + q_2 = 1$. The following questions are of practical interest.

With initial condition $Y_m(0) = x_m$, what are asymptotically (as $m \to \infty$):

- the hitting place (Z_m) distribution on the (A) boundary?

- the hitting time (T_m) distribution on the (A) boundary?

The covariance matrix of one step is given by

$$CY = \begin{pmatrix} p_1 + q_1 - (p_1 - q_1)^2 & -(p_1 - q_1)(p_2 - q_2) \\ -(p_1 - q_1)(p_2 - q_2) & p_2 + q_2 - (p_2 - q_2)^2 \end{pmatrix} \tag{3.53}$$

If we let the drift (or trend) μ be $\mu = (\mu_1, \mu_2) = (p_1 - q_1, p_2 - q_2)$, we have two fundamental different limit behaviors, according to $\mu = 0$ or $\mu \neq 0$

7.1.1   THE TREND-FREE CASE: $\mu = 0$

To simplify the analysis, assume that $p_1 = q_1 = p_2 = q_2 = 1/4$. The asymptotic distribution of $Y_m(.)$ is given by the following theorem

THEOREM 3.18

$$\frac{Y_m(2[m^2t])}{m} \Rightarrow_{m \to \infty} W(t) \tag{3.54}$$

where $W(t)$ is a two-dimensional reflected and absorbed Brownian motion with boundary conditions similar to these of Figure 3.9. Let T be the hitting time for $W(.)$. The density of $W(.)$ is given by

$$P_x(W(t) \in dy, t < T) = \{2 \sum_{k=1}^{\infty} exp(-k^2\pi^2t/2)\cos(k\pi x_1)[(\cos(k\pi y_1)$$

$$- \cos(h\pi(1-y_2))]] + 2\sum_{k=1}^{\infty} exp(-l^2\pi^2t/2)\cos(l\pi x_2)$$

$$\times [\cos(l\pi y_2) - \cos[h\pi(1-y_1)]]$$

$$+ 4\sum_{k=1}^{\infty}\sum_{l=1}^{\infty} exp[-(k^2+l^2)\pi^2t/2)]\cos(k\pi x_1)\cos(l\pi x_2)[\cos(k\pi y_1)\cos(l\pi y_2)$$

$$- \cos[k\pi(1-y_2)]\cos[l\pi(1-y_1)]]\}dy_1 dy_2$$

$$\tag{3.55}$$

Proof: The weak convergence of Eq.(3.54) is easily deduced from [Chung and Williams, 1983]. We skip the proof of this part since it is not very interesting for the reader. See [Louchard and Schott, 1991] for details.

The factor 2 is derived from σ_y^2 as given by Eq.(3.53). The reflecting properties are such that, prior to absorption (in one dimension, for instance),

$$Y_m(n) = X(n) + max_{0 \leq i \leq n}(-X(i))$$

where

$$X(n) = \sum_{i=1}^{n} \Delta_i$$

and

$$P(\Delta_i = 1) = p, P(\Delta_i = -1) = 1 - p = q$$

The hitting time T is (almost surely) a continuous functional of W [Chung and Williams, 1983] so that the weak convergence is also valid for our absorbed process. Let W_1 be the one-dimensional Brownian motion (9.1) with reflecting boundaries at 0 and 1. Its density is given by [Feller, 1970]:

$$
\begin{aligned}
P_{x_1}(W_1(t) \in dy_1) &= \{1 + 2\sum_{k=1}^{\infty} exp(-k^2\pi^2 t/2)\cos(k\pi x_1)cos(k\pi y_1)\}dy_1 \\
&= (1 + A_1(x_1, y_1))dy_1
\end{aligned}
$$

and similarly for W_2.

Now we apply the reflection principle across the absorbing boundary:

$$
\begin{aligned}
P_x(W(t) \in dy, t < T) &= \{(1 + A_1(x_1, y_1))(1 + A_2(x_2, y_2)) \\
&\quad - (1 + A_1(x_1, 1 - y_2))(1 + A_2(x_2, 1 - y_1))\}dy_1 dy_2
\end{aligned}
$$

hence we get Eq.(3.55) We have now all necessary tools to derive the

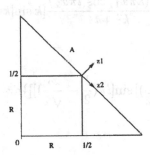

Figure 3.10. New coordinate system

limiting hitting distributions. Let a new coordinate system z be given by the translation and rotation from the origin O to the middle point of the absorbing boundary as shown on Figure 3.10.

We have the following result for the distributions of hitting place Z and hitting time T on the diagonal.

THEOREM 3.19

$$P_x[T \in dt, \ Z(T) \in dz_2] =$$

$$\{\sum_{k=1}^{\infty} exp(-\frac{k^2\pi^2}{2}\cos(k\pi x_1)\frac{2k\pi}{\sqrt{2}}\sin[k\pi(\frac{1}{2}+\frac{z_2}{\sqrt{2}}]$$

$$+\{\sum_{l=1}^{\infty} exp(-\frac{l^2\pi^2}{2}\cos(l\pi x_2)\frac{2l\pi}{\sqrt{2}}\sin[l\pi(\frac{1}{2}-\frac{z_2}{\sqrt{2}})]$$

$$+\sum_{k=1}^{\infty}\sum_{l=1}^{\infty} exp[-(k^2+l^2)\pi^2\frac{t}{2}]\cos(k\pi x_1)\cos(l\pi x_2)$$

$$\times \frac{4\pi}{\sqrt{2}}[ksin[k\pi(\frac{1}{2}+\frac{z_2}{\sqrt{2}})]\cos[l\pi(\frac{1}{2}-\frac{z_2}{\sqrt{2}})]$$

$$+l\cos[k\pi(\frac{1}{2}+\frac{z_2}{\sqrt{2}})]\sin[l\pi(\frac{1}{2}-\frac{z_2}{\sqrt{2}})]]\}dtdz_2$$

(3.56)

The marginal densities are given by:

$$P_x[Z(T) \in dz_2] =$$

$$\{\frac{4}{\pi\sqrt{2}}[\sum_{k=1}^{\infty}\frac{\cos(k\pi x_1)}{k}\sin[k\pi(\frac{1}{2}+\frac{z_2}{\sqrt{2}})]$$

$$+\sum_{k=1}^{\infty}\frac{\cos(l\pi x_2)}{l}\sin[l\pi(\frac{1}{2}-\frac{z_2}{\sqrt{2}})]$$

$$+\frac{8}{\pi\sqrt{2}}\sum_{k=1}^{\infty}\sum_{l=1}^{\infty}\frac{\cos(k\pi x_1)\cos(l\pi x_2)}{k^2+l^2}[k\sin[k\pi(\frac{1}{2}+\frac{z_2}{\sqrt{2}})]$$

$$\times \cos[l\pi(\frac{1}{2}-\frac{z_2}{\sqrt{2}})]$$

$$+[l\cos[k\pi(\frac{1}{2}+\frac{z_2}{\sqrt{2}})]\sin[l\pi(\frac{1}{2}-\frac{z_2}{\sqrt{2}})]]\}dz_2$$

(3.57)

$$P_x[T \in dt] =$$

$$\{4\sum_{k \ odd>0} exp(-k^2\pi^2\frac{t}{2})\cos(k\pi x_1)+4\sum_{l \ odd>0} exp[-\frac{l^2\pi^2}{2}\cos(l\pi x_2)$$

$$+8\sum_{k \ odd>0}\sum_{l \ even>0} exp[-(k^2+l^2)\pi^2\frac{t}{2}]\frac{(k^2+l^2)}{(k^2-l^2)}\cos(k\pi x_1)\cos(l\pi x_2)$$

$$-8\sum_{keven>0}\sum_{lodd>0} exp[-(k^2+l^2)\pi^2\frac{t}{2}]\frac{(k^2+l^2)}{k^2-l^2}\cos(k\pi x_1)\cos(l\pi x_2)\}dt$$

(3.58)

Proof: The new coordinate system (z_1, z_2) is given by

$$\begin{cases} z_1 = \frac{-\sqrt{2}}{2}(1 - y_1 - y_2) \\ z_2 = \frac{\sqrt{2}}{2}(y_1 - y_2) \end{cases}$$

In this new system let

$$\phi_x(t, z_1, z_2) = P_x[W(t) \in dz, t < T]$$

The hitting density is given by:

$$P_x[T \in dt, Z(t) \in dz_2] = -\frac{1}{2}[\partial_{z_1} \phi_x(t, z_1, z_2)]_{z_1 = 0}$$

This gives Eq.(3.55) after standard computations. Integrating on t gives Eq.(3.57). Eq.(3.58) is obtained from Eq.(3.56) after some tedious but simple computations and one checks that Eq.(3.58) is a probability density.

By Eq.(3.54) the hitting place Z_m and hitting time T_m are given by:

$$\frac{Z_m}{m} \sim_D Z$$

and

$$\frac{T_m}{2m^2} \sim_D Z$$

Z and T are almost surely continuous functionals of W. The case $p_1 = q_1$, $p_2 = q_2$ with $p_1 \neq p_2$ is also trend free but the reflection principle is no more available.

The case $\mu \neq 0$ is more intricate. See [Louchard and Schott, 1991] for details.

7.2   BANKER ALGORITHM

In this section we analyse the restricted version of this algorithm which involves only two customers C_1 and C_2 sharing a fixed quantity of a given resource M (money). There are fixed upper bounds m_1 and m_2 on how much of the resource each of the customers is allowed to use at any time. The banker decides to affect to the customer C_i, $i = 1, 2$ the required resource units only if the remaining units are sufficient in order to fulfill the requirements of C_j, $j = 1, 2; j \neq i$. This situation is modelled (see Figure 3.11) by a random walk in a rectangle with a broken corner , i.e.,

$$\{(x_1, x_2) : 0 \leq x_1 \leq m_1, 0 \leq x_2 \leq m_2, x_1 + x_2 \leq m\}$$

where the last constraint generates the broken corner. The random walk is reflected on the sides parallel to the axes and is absorbed on the sloping side.

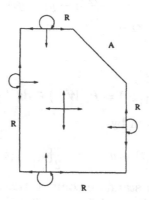

Figure 3.11.   Banker algorithm

To simplify the analysis, we assume that $m_1 = m_2 = m$. As for the two stacks problem several subcases must be considered. It appears that the trend-free case $\mu = 0$ is the most intricate (see [Louchard and Schott, 1991, Louchard et al., 1994] for a complete analysis). If $\mu \neq 0$ the analysis proceeds exactly as previously and we get, for example the following result:

THEOREM 3.20 *Assume that: $p_1 = p_2 = p$, $q_1 = q_2 = q$, $p < q$. Then the hitting place Z_m is asymptotically uniformly distributed on the absorbing boundary. ϵT_m is asymptotically distributed as an exponential random variable, where ϵ is a constant depending only on m, p and q.*

Proof: For a two reflecting barriers, one-dimensional random walk, the stationary distribution (cf. page 12) is again reached with exponential speed (we just have to change $P(Y_1 = 0)$ by suitable normalization. If we ignore the absorbing boundary in our problem, the mean return time to 0 is given by ([Louchard and Schott, 1991], Theorem 2.12) . For $j \leq m$, the hitting probability satisfies the same equations as in the proof of ([Louchard and Schott, 1991], Theorem 2.10), so that (with this proof's notation)

$$\phi_j^* = 1 + y_0 \Psi(j - 1)$$

and

$$\phi_m^* = \phi_{m-1}^* + \frac{y_0}{m}\left(\frac{q}{p}\right)^{m-1}$$

At m, we have, by our random walk reflecting properties

$$\phi_m^* = (\frac{m}{m+1})[p'\phi_{m+1}^* + q'\phi_{m-1}^*] + \frac{1}{m+1}\phi_m^*$$

which yields

$$\Delta_{m+1} = \phi_{m+1}^* - \phi_m^* = (\frac{y_0}{m})(\frac{q}{p})^m$$

For $j > m$, the equations are

$$\phi_j^* = q'\phi_{j-1}^* + \frac{p'}{2m-j+1}\phi_j^* + p'\frac{2m-j}{2m-j+1}\phi_{j+1}^*$$

and, after some difference equation manipulation,

$$\phi_j^* = \phi_m^* + y_0\bar{\Psi}(j-1)$$

where

$$\bar{\Psi}(j) = \sum_{i=m}^{j}\frac{(\frac{q}{p})^i}{2m-i}$$

We get

$$\bar{\Psi}(j-1) \sim \frac{(\frac{q}{p})^j}{(2m-j+1)(\frac{q}{p}-1)}[1+O(\frac{1}{2m-j})]$$

The condition $\phi_{m'}^* = 0$ now leads to $y_0 \sim -(2m-m')(\frac{q}{p}-1)(\frac{p}{q})^m$ and we finally obtain $\phi_1^* \sim 1 - (2m-m')(\frac{p}{q})^m$. We now follow the proof of ([Louchard and Schott, 1991], Theorem 2.13) to get our result.

7.3   A LARGE DEVIATION ANALYSIS OF THE TWO STACKS PROBLEM

To illustrate the use of large deviations theory we analyse the two stacks problem as we did for dynamic algorithms in Section 2.4.2. The results of this section are due to R. Maier [Maier, 1991b]. Here we allow some space dependency for the transition probabilities.

7.3.1   A ONE-STACK MODEL

Consider a single stack evolving in a region of size m. The stack begins empty and is subjected to a random sequence of insertions (I) and deletions (D). When the stack height equals j the next operation will be chosen randomly, with probabilities

$$P(a) = \begin{cases} \frac{1}{2} - \frac{1}{2}g(\frac{j}{m}) & a = I \\ \frac{1}{2} + \frac{1}{2}g(\frac{j}{m}) & a = D \end{cases}$$

where $g(.) \geq 0$ is some sufficiently smooth function defined on $[0, 1]$. $g(x)$ measures the extent to which deletions from a stack predominate over insertions, as a fraction of x, the fraction of available memory occupied by the stack. The traditional model has $g(x) = 1 - 4p$, a constant function of x. The state space is

$$Q_m = \{j \in Z | 0 \leq j \leq m\}$$

The random variable Δ, the amount by which the stack height changes under the action of a single operation, has its distribution parametrised by the normalised stack height $x = \frac{j}{m}$. In fact

$$P(\Delta(x) = v) = \begin{cases} \frac{1}{2} - \frac{1}{2}g(x) & v = 1 \\ \frac{1}{2} + \frac{1}{2}g(x) & v = -1 \end{cases}$$

x may be regarded as a function of t, the operation count divided by m; its initial value is $x_0 = 0$. In any event, suppose that the stack height is not to exceed l, some fixed fraction of m. The stack height will reach l when $x(t)$ reaches $x_1 = \frac{l}{m}$. So this is a one-dimensional exit time problem. The region B is $[0, x_1]$ and we can apply directly Ventcel and Freidlin's exit time results [Freidlin and Wentzell, 1984].

THEOREM 3.21 *[Maier, 1991b] As $m \to \infty$, the random variable t_{exit} is asymptically exponential. Its expectation satisfies*

$$lim_{m \to \infty} \frac{1}{m} log E(t_{exit}) = S_0$$

in which S_0 is the instanton action. That is to say S_0 is the asymptotic exponential growth rate of $E(t_{exit})$ as $m \to \infty$.

Proof: This is a general result of Ventcell and Freidlin's theory. We skip the proof (see [Freidlin and Wentzell, 1984, Maier, 1991b].

The generating function of Δ is

$$E(e^{p\Delta(x)}) = \cosh p - g(x) sinh p$$

and the associated one-stack Hamiltonian is

$$H^{(1)}(x, p) = \log E(e^{p\Delta(x)}) = \log[\cosh p - g(x) \sinh p] \qquad (3.59)$$

The zero-energy constraint $E(e^{p\Delta(x)}) = 1$ determines $p(x)$ for the instanton path. The equation $cosh\, p - g(x)\sinh p = 1$ has two solutions; one is $p = 0$, and the other is

$$p(x) = 2tanh^{-1}g(x)$$

The instanton has action

$$S_0 = \int_{x_0}^{x_1} p(x)dx = 2\int_0^{x_1} \tanh^{-1}g(x)dx$$

and the following result follows from Theorem 3.21

THEOREM 3.22 *Asymptotically, the number of operations that take place in the single-stack model before stack height reaches $x_1 m$ is an exponential random variable. Its mean has asymptotics*

$$E(\tau) = \Theta_{\log}(e^{2m\int_0^x \tanh^{-1} g(x)dx}),\ \text{as}\ m \to \infty.$$

7.3.2   A TWO-STACK MODEL

Consider now the two-stack model defined in Section 7.1. The normalized state space Q is the triangular region

$$Q = \{(x,y) \in \mathcal{R}^2 | x \geq 0, y \geq 0, x + y \leq 1\}$$

and the processes $x^{(}m)(t)$ represent the normalised state $(j/m, k/m)$ as a function of t, the operation count divided by m. B, the region from which the $x^{(}m)(t)$ attempt to exit is Q itself. The jumps in $x^m(t)$ are of size $m^{-1}\Delta$, with $\Delta(x)$ distributed according to

$$
\begin{aligned}
\bar{\alpha} &= 1 + E[\text{Number of transitions to state 2} \mid \text{The chain is in state 1}] \\
&= \sum_{i=0}^{\infty} iq^i(1-q) \\
&= \frac{1}{1-q} - 1 \\
&= \frac{q}{1-q}
\end{aligned}
$$

$$
\begin{aligned}
P(\Delta(x) = v) &= \frac{1}{4} - \frac{1}{4}g(x), v = (1,0) & (3.60) \\
&= \frac{1}{4} + \frac{1}{4}g(x), v = (-1,0) & (3.61) \\
&= \frac{1}{4} - \frac{1}{4}g(y), v = (0,1) & (3.62) \\
&= \frac{1}{4} + \frac{1}{4}g(y), v = (0,-1). & (3.63)
\end{aligned}
$$

The exit time is the time when the stacks collide (i.e. when $x(t)$ exits Q through the line $x + y = 1$, assuming that x_0 equalled $(0,0)$).
The generating function of $\Delta(x)$ follows from Eq.(3.63). It is:

$$E(e^{p \cdot \Delta(x)}) = \frac{1}{2} \cosh p_x + \frac{1}{2} \cosh p_y - \frac{1}{2} g(x) \sinh p_x - \frac{1}{2} \sinh p_y \quad (3.64)$$

Since

$$
\begin{aligned}
H(x,p) &= \log E(e^{p \cdot \Delta(x)}) \\
&= -\log(2) + \log[\cosh p_x - g(x) \sinh p_x + \cosh p_y - g(y) \sinh p_y]
\end{aligned}
$$

one has

$$H(x,p) = log(\frac{e^{H^{(1)}(x,p_x)} + e^{H^{(1)}(y,p_y)}}{2})$$

In which $H^{(1)}(.,.)$ is the single-stack Hamiltonian defined in Eq.(3.59). The following result from Hamiltonian dynamics applies:

PROPOSITION 5 *Suppose that $H(x,p)$ is an explicit function of the one-dimensional Hamiltonians $H^{(1)}(x,p_x)$ and $H^{(1)}(y,p_y)$. Then E_x and E_y, the values of these one-dimensional Hamiltonians, will be constants of the motion. If $x_{E_x(t)}$ and $x_{E_y(t)}$ are trajectories of energy E_x and E_y arising from $H^{(1)}$, then*

$$x(t) = (x_{E_x}(\frac{\partial H}{\partial E_x}(E_x, E_y)t), x_{E_y}(\frac{\partial H}{\partial E_y}(E_x, E_y)t))$$

will be a solution of the two-dimensional equations of motion. Moreover, all solutions will be of this form. A similar decomposition

$$p(t) = (p_{E_x}(\frac{\partial H}{\partial E_x}(E_x, E_y)t), p_{E_y}(\frac{\partial H}{\partial E_y}(E_x, E_y)t))$$

holds for the momentum $p(t)$.

In our case

$$H(E_x, E_y) = log(\frac{e^{E_x} + e^{E_y}}{2}) \quad (3.65)$$

therefore:

$$x(t) = (x_{E_x})(\frac{e^{E_x}}{e^{E_x} + e^{E_y}}t, x_{E_y}(\frac{e^{E_y}}{e^{E_x} + e^{E_y}}t)) \quad (3.66)$$

and

$$p(t) = (p_{E_x}(\frac{e^{E_x}}{e^{E_x} + e^{E_y}t}), p_{E_y}(\frac{e^{E_y}}{e^{E_x} + e^{E_y}}t))$$

The two-stack problem thus separates into two coupled one-stack problems.

According to Proposition 5, the trajectory of the two-stack instanton $x^*(t)$ has zero energy. So $H(x,p) = 0$, and by Eq.(3.65)

$$e^{E_x} + e^{E_y} = 2 \qquad (3.67)$$

This constraint does not uniquely determine $x^*(t)$. There are in general an infinite number of solutions $x(.)$ satisfying both Eq.(3.67) and the boundary condition $x(0) = (0,0)$.

Suppose $g(0) = 0$, so that insertions into and deletions from each stack are equiprobable at low stack height. In that case the equations $H^{(1)}(x,p_x) = E_x$ and $H^{(1)}(y,p_y) = E_y$ i.e.,

$$\log[\cosh p_x - g(x)\sinh p_x] = E_x, \log[\cosh p_y - g(y)\sinh p_y] = E_y$$

are consistent with Eq.(3.67) at $(x,y) = (0,0)$ only if $E_x = E_y = 0$. Substitution of $E_x = E_y = 0$ in Eq.(3.66) yields

$$(x_{E_x=0}(t/2), x_{E_y=0}(t/2))$$

7.4   EXHAUSTION OF SHARED MEMORY: GENERAL RESULTS

The material of this section is taken from [Maier and Schott, 1993a]. As shown in previous sections, in simple models of dynamic data structures it is possible to investigate the memory exhaustion time. There has however been very little work on the exhaustion of *shared* memory, or on 'multidimensional' exhaustion, where one of a number of inequivalent resources becomes exhausted.

In this section we go beyond previous work by studying the exhaustion of shared storage: we consider the interaction of q independent processes $P_1, \ldots, P_q$, each with its own memory needs. We allow the processes to allocate and deallocate r different, non-substitutable resources (types of memory): $R_1, \ldots, R_r$. We model resource limitations, and define resource exhaustion, as follows. At any time s, process P_i is assumed to have allocated some quantity $y_i^j(s)$ of resource R_j. (Both time and resource usage are taken to be discrete, so that $s \in \mathcal{N}$ and $y_i^j(s) \in \mathcal{N}$.) Process P_i is assumed to have some maximum need m_{ij} of resource R_j, so that

$$0 \leq y_i^j(s) \leq m_{ij} \qquad (3.68)$$

for all s. m_{ij} may be infinite; if finite, it is a hard limit which the process P_i never attempts to exceed. The resources R_j are limited, so that

$$\sum_{i=1}^{q} y_i^j(s) < m_j \qquad (3.69)$$

for $m_j - 1$ the total amount of resource R_j available for allocation. Resource exhaustion occurs when some process P_i issues an unfulfillable request for a quantity of some resource R_j. Here 'unfulfillable' means that fulfilling the request would violate one of the inequalities (3.69).

The state space Q of the memory allocation system is the subset of $\mathcal{N}^{qr}$ determined by (3.68) and (3.69). This polyhedral state space is familiar: it is used in the banker's algorithm for deadlock avoidance. However most treatments of deadlocks (see Habermann [Habermann, 1978]) assume that processes request and release resources in a mechanical way: a process P_i requests increasing amounts of each resource R_j until the corresponding goal m_{ij} is reached, then releases resource units until $y_i^j = 0$, and repeats. (The r different goals of the process need not be reached simultaneously, of course.) This is a powerful assumption: it facilitates a classification of system states into 'safe' and 'unsafe' states, the latter being those which can lead to deadlock. However it is an idealization.

We analyse here an altogether different stochastic model, one more suited to the memory usage of dynamic data structures. We assume that regardless of the system state, each process P_i with $0 < y_i^j < m_{ij}$ can issue either an allocation or deallocation request for resource R_j. The probabilities of the different sorts of request may depend on the current state vector (y_i^j). In other words we take the state of the storage allocation system as a function of time to be a finite-state Markov chain; this is an alternative approach which goes at least as far back as Ellis [Ellis, 1977].

Remember that the goal is the estimation of the amount of time τ until memory exhaustion occurs, if initially the r types of resources are completely unallocated: $y_i^j = 0$ for all i, j. We are particularly interested in *asymptotics*: the consequences of expanding the resource limits $m_j - 1$ and the per-process maximum needs m_{ij} (if finite) on the expected time to exhaustion. For this we must specify the stochastic model more precisely. First we make the realistic assumption that at each time $s \in \mathcal{N}$

and for each pair (i, j), the change in the usage of resource R_j by process P_i due to allocations or deallocations has a *negative expectation*, provided of course that $y_i^j > 0$. (The less realistic case of positive expectation has been considered by Yao [Yao, 1981].) Under this assumption the Markov chain manifests a sort of stability: the initial state is an equilibrium state, although large fluctuations away from it occasionally occur, and eventually end in memory exhaustion. We also assume that the q processes behave identically, and that their needs for the r different resources are equal. These last assumptions make the mathematics symmetric: for all i, j we have $m_i = m'$ and $m_{ij} = m''$ for some m', m''.

At each time $s \in \mathcal{N}$ the state vector y_i^j will change by ξ_i^j, an independent instance of a $\mathcal{Z}^{qr}$-valued random vector. There is probability $(qr)^{-1}$ of any process $P_{i'}$ issuing a request concerning any resource type $R_{j'}$, on account of symmetry. If this occurs, the random vector ξ_i^j will equal $\xi \delta_{i,i'} \delta_{j,j'}$, in which ξ is a $\mathcal{Z}$-valued random variable signifying the requested change in resource allocation. We allow the distribution of ξ to depend on the current state as follows: it is some specified function of $y_{i'}^{j'}/m'$, the amount of resource $R_{j'}$ currently allocated to process $P_{i'}$ as a fraction of the total amount available. This yields a very general but reasonable assumption: each process evolves independently, and its behavior *vis-à-vis* some resource (the probabilities of its issuing allocation and deallocation requests for the resource) depends only on that fraction of the resource which it has currently allocated.

Actually the preceding description of the distribution of ξ_i^j can hold only in the interior of Q; near the boundary the transition probabilities must differ. For example if $y_i^j = m''$, then ξ_i^j cannot be positive. In effect there must be 'reflecting boundary conditions' on the faces of Q determined by (3.68), just as there are final states, or 'absorbing boundary conditions' on the exhaustion faces determined by (3.69).

We shall answer the following 'scaling up' question: if for certain positive α' and α'' we take $m' = \lfloor N\alpha' \rfloor$, $m'' = \lfloor N\alpha'' \rfloor$, what are the large-$N$ asymptotics of $E(\tau)$, the expected time until exhaustion of one of the resources occurs? Scaling up signifies increasing both resource limits and (if present) per-process maximum resource needs, but keeping process *dynamics* fixed: the distribution of ξ, the random variable signifying a requested change in resource usage, is independent of the scaling parameter N.

A very special case of our model would set

$$P(\xi = k) = \begin{cases} p, & k = 1 \\ 1 - p, & k = -1 \end{cases} \qquad (3.70)$$

for some fixed $p < \frac{1}{2}$. With this choice, p would be the probability of an allocation request (for one resource unit) and $1 - p$ the probability of a deallocation. The model defined by (3.70), with two processes and one resource, has a long history. Knuth ([Knuth, 1973], Ex. 2.2.13) proposed a model very similar to it: a model of *colliding stacks* in which two stacks (heights y_1^1 and y_2^1) are allowed to grow from opposite sides of a fixed block of memory, and memory exhaustion occurs when the stacks collide. In his model m'' was effectively infinite, so the state space Q was a triangle in $\mathcal{N}^2$, with two reflecting edges ($y_1^1 = 0$, $y_2^1 = 0$) and a memory exhaustion edge ($y_1^1 + y_2^1 = m'$, the total memory available).

That $E(\tau)$ grows exponentially in N is intuitively obvious, but the uniform distribution over exhaustion states is counterintuitive. Maier [Maier, 1991b] showed that this distribution is an artifact: if for each process the probability p of issuing an allocation request is allowed to depend on the fraction of memory currently allocated, very different behavior may obtain. If p is a *decreasing* function of this fraction (so that the model is 'increasingly contractive', with large fluctuations away from the initial state strongly suppressed), then the final state will be concentrated close to the exhaustion face of Q.

Below we apply the asymptotic techniques of Matkowsky, Schuss and coworkers [Knessl et al., 1985, Naeh et al., 1990] to the general model: q and r arbitrary, and the distribution of the increment ξ_i^j, for any pair (i, j), allowed to depend in some universal way on y_i^j/m''; equivalently, on y_i^j/N. The technique introduced in Ref. [Naeh et al., 1990] allows a *complete* determination of the asymptotics of the mean exhaustion time $E(\tau)$: not merely the rate of exponential growth with N, but also the pre-exponential factor. The problem of general q and r was left open in Ref. [Louchard and Schott, 1991], and is now largely solved. The results are summarized in the sequence of theorems beginning with Theorem 3.24. We obtain precise large-N asymptotics of $E(\tau)$ if p is constant, and somewhat less precise asymptotics for the increasingly contractive model.

7.4.1   EXIT TIME THEORY

We summarize the ideas of Matkowsky, Schuss *et al.*, adapted to the case of Markov chains. (Their earlier paper Ref. [Knessl et al., 1985] concentrated on Markov chains; their recent key paper Ref. [Naeh et al., 1990] deals with the exit problem for continuous-time Markov processes.) Their technique of 'matched asymptotic expansions' is precisely what is needed to obtain the large-N asymptotics of $E(\tau)$.

Let $Q \subset \mathcal{R}^d$ be a *normalized state space*: we assume that Q is closed, connected and bounded with a sufficiently well-behaved boundary, and that $0 \in Q$. Define $Q \subset Z^d$, an N-dependent state space, to be the set of all d-tuples of integers $(y_1, \ldots, y_d)$ such that $N^{-1}(y_1, \ldots, y_d) \in Q$. We introduce a Z^d-valued increment random variable ξ, whose distribution is parametrized by the current normalized state $\mathbf{x} \in Q$. A Markov chain $\mathbf{y}(0), \mathbf{y}(1), \ldots$ on Q is defined by specifying that any time s, $\Delta \mathbf{y}(s) \stackrel{\text{def}}{=} \mathbf{y}(s+1) - \mathbf{y}(s)$ has the same distribution as $\xi(\mathbf{y}/N)$. Since we want the state $\mathbf{0}$ to be a point of stable equilibrium, we require that the 'mean drift' vector field $E(\xi(\cdot))$ on Q have $\mathbf{0}$ as a stable fixed point.

We must specify appropriate boundary conditions for this Markov chain. Some parts of the boundary of Q, which we denote ∂Q, are taken to be 'absorbing.' The corresponding states in Q are viewed as final states; whenever the system enters such a state, its evolution terminates. Also, some portions of the boundary may be 'reflecting': the transition probabilities of the chain, i.e., the distribution of ξ, are modified there so as to prevent the normalized state from leaving Q.

For any N, Q is a finite set. The transition matrix of the Markov chain on Q, $\mathbf{T} = (T_{\mathbf{yz}}) \in \mathcal{R}_+^{Q \times Q}$, is a substochastic matrix whose elements follow from the distribution of ξ. It is strictly substochastic (i.e., $\sum_{\mathbf{z} \in Q} T_{\mathbf{yz}} \leq 1$ for all $\mathbf{y}$ rather than $\sum_{\mathbf{z} \in Q} T_{\mathbf{yz}} = 1$ for all $\mathbf{y}$) on account of the absorption on the boundary. By standard Perron-Frobenius theory [Minc, 1988], $\mathbf{T}$ has a left eigenvalue $\lambda_1 > 0$ of maximum modulus, and a corresponding positive left eigenvector $\rho = (\rho_{\mathbf{y}}) \in \mathcal{R}_+^Q$. If $\mathbf{T}$ were stochastic, this eigenvector would be interpreted as the (unnormalized) stationary density of the chain, and the eigenvalue would equal 1. But since it is not, $\lambda_1 < 1$ and ρ is interpreted as a 'quasi-stationary' density: the distribution of the system state over Q, if one conditions on the event that after a very long time, absorption on the boundary has failed to occur. (One expects [Naeh et al., 1990] that ρ is strongly concentrated near the point of stable equilibrium $\mathbf{0}$, much like a stationary density.) $1 - \lambda_1$ is a *limiting* (or *quasi-stationary*) *absorption rate*: we shall see that it falls to zero exponentially in N. Other eigenvalues of $\mathbf{T}$, it turns out, are well separated from λ_1 as $N \to \infty$, so we are justified in approximating the mean absorption time $E(\tau)$ by $(1 - \lambda_1)^{-1}$ in the large-N limit. The technique of Ref. [Naeh et al., 1990] is really a technique for computing the large-N asymptotics of $1 - \lambda_1$.

The idea is to approximate the quasi-stationary eigenvector $\rho_{\mathbf{y}}$, when N is large, in three different regions: (1) near $\mathbf{y} = \mathbf{0}$, where most of the probability is concentrated, (2) along certain trajectories between $\mathbf{0}$ and the absorbing boundary, and (3) in a neighborhood of the absorbing boundary. By matching these three expressions together a consistent set of approximations to ρ is obtained. Surprisingly, to leading order in N (as $N \to \infty$) it is possible to do all this without knowing the exponentially small quantity $1 - \lambda_1$: in effect, λ_1 may be taken equal to unity throughout. After ρ is sufficiently well approximated, the large-N asymptotics of $1 - \lambda_1$ are computed as an asymptotic absorption rate:

$$1 - \lambda_1 \sim \frac{\sum_{y \in Q}(1 - \sum_{z \in Q} T_{yz})\rho_z}{\sum_{y \in Q} \rho_y}. \qquad (3.71)$$

Here $(1 - \sum_{z \in Q} T_{yz})\rho_y$ is the amount of probability absorbed at state y per time step, on account of the substochasticity; it is nonzero only for y sufficiently near the absorbing boundary. The numerator will typically be much smaller than the denominator (which is simply a normalization factor) because ρ_y, for y near the absorbing boundary, will when N is large be exponentially smaller than ρ_y, for y near the point of stable equilibrium. In fact the denominator may be computed from the approximation to ρ_y in Region 1 alone. The numerator is computed from the approximation to ρ_y in Region 3, and the approximation in Region 2 merely serves to ensure consistency between the approximations in Regions 1 and 3.

7.4.2   THE $Q = 1$, $R = 1$ MODEL

We first apply the approach of the last section to a model with dimensionality $d = 1$: the $q = 1$, $r = 1$ case of the memory exhaustion model of Section 8.1. This is a model of a single process P_1 and a single resource R_1 available in limited quantities. Let $\mathcal{Q} = [0, \alpha]$, so that $Q = \{0, \dots, \lfloor N\alpha \rfloor\}$ is the state space. (There is no need to distinguish between α' and α'' here.) Let the random increment variable $\xi = \Delta y$ have density given by (3.70), but with p allowed to be a function of the normalized state $x \overset{\text{def}}{=} y/N$. (The process will allocate and deallocate resource units singly, with the respective probabilities allowed to depend on the fraction of the resource which is currently allocated.) We assume $0 < p(x) < \frac{1}{2}$ for all $x \in \mathcal{Q}$, so the state 0 (the initial state) is stable. The state $\lfloor N\alpha \rfloor$ is the only final state.

The transition probabilities T_{yz} must be modified when $y = 0$ or $y = \lfloor N\alpha \rfloor - 1$. The state 0 is reflecting; we take $T_{00} = 1 - p_0$, $T_{01} = p_0$, for some parameter $p_0 \in (0, 1]$ which specifies how 'hard' the reflection is. (In the computing context p_0 is the probability that the process P_1, if the resource has been completely deallocated, will immediately issue an allocation request.) And since the state $\lfloor N\alpha \rfloor$ is final and probability is 'lost' at state $\lfloor N\alpha \rfloor - 1$, we set $T_{\lfloor N\alpha \rfloor - 1, \lfloor N\alpha \rfloor} = 0$. With these choices the model becomes identical to the one-stack model of Maier [Maier, 1991b].

Consider the quasi-stationary density ρ, approximated as the solution of $\rho\mathbf{T} = \rho$, in Region 1. In this region $y = o(N)$, *i.e.*, $x = o(1)$. To leading order $P(\Delta y = +1) = p(0)$ and $P(\Delta y = -1) = 1 - p(0)$; it is clear that

$$\rho_y = \begin{cases} [p(0)/(1 - p(0))]^y, & y > 0 \\ p(0)/p_0, & y = 0 \end{cases} \tag{3.72}$$

satisfies $\rho\mathbf{T} = \rho$. The parameter p_0 affects ρ_0 only.

The behavior of ρ_y in Region 3 is similar. Here $\lfloor N\alpha \rfloor - y = o(N)$, *i.e.*, $\alpha - x = o(1)$. As in Region 1, transition probabilities are essentially constant: $P(\Delta y = +1) = p(\alpha)$ and $P(\Delta y = -1) = 1 - p(\alpha)$. We set $\rho_{\lfloor N\alpha \rfloor} = 0$ since probability, once absorbed in the final state, never returns. The approximation

$$\rho_y = C_3 \left([p(\alpha)/(1 - p(\alpha))]^y - [p(\alpha)/(1 - p(\alpha))]^{\lfloor N\alpha \rfloor} \right) \tag{3.73}$$

will satisfy $\rho\mathbf{T} = \rho$. C_3 will be chosen to ensure consistency between (3.72) and (3.73). To ensure consistency between the approximations to ρ in Regions 1 and 3, we use a special sort of approximation: in the intermediate Region 2. There we take

$$\rho_y = K(y/N) \exp\left(-NW(y/N)\right). \tag{3.74}$$

This sort of approximate, but asymptotically correct, solution to an eigenvector (or eigenfunction) equation is traditionally called a 'WKB solution' ([Naeh et al., 1990]; see Section 3 of Ref. [Knessl et al., 1985] for the Markov chain case). The two functions $K(\cdot)$ and $W(\cdot)$ on Q are determined by the x-dependence of the density of ξ. With the choice (3.70), it suffices to take

$$W(x) = \int_0^x \log[(1 - p(z))/p(z)] \, dz \tag{3.75}$$

$$K(x) = C_2[p(x)(1 - p(x))]^{-1/2} \tag{3.76}$$

with C_2 some constant. It is an elementary, though tedious, exercise to verify that with these choices for $W(\cdot)$ and $K(\cdot)$ we have $\rho\mathbf{T} = \rho$, *i.e.*,

$$\rho_{y-1}p((y - 1)/N) + \rho_{y+1}(1 - p((y + 1)/N)) = \rho_y \tag{3.77}$$

to leading order in N as $N \to \infty$.

(For a general $d = 1$ model, with ξ not restricted to ± 1 values, the formulæ for $W(x)$ and $K(x)$ would differ; in general $W(x) = \int_0^x p^*(z) \, dz$, with $p^* = p^*(x)$ defined as the positive solution to the implicit equation $\mathbf{E}e^{p^*\xi(x)} = 1$. See Maier [Maier, 1991b] for more.)

The WKB solution (3.74) is of exactly the form needed to interpolate between (3.72) and (3.73). It has asymptotics

$$\rho_y \sim C_2[p(0)(1 - p(0))]^{-1/2}[p(0)/(1 - p(0))]^y, \quad y = o(N) \qquad (3.78)$$

$$\rho_y \sim C_2[p(\alpha)(1 - p(\alpha))]^{-1/2}e^{-NW(\alpha)}[p(\alpha)/(1 - p(\alpha))]^{y-\lfloor N\alpha\rfloor}, \qquad (3.79)$$
$$y = \lfloor N\alpha\rfloor - o(N).$$

Equating coefficients in (3.72) and (3.78) gives $C_2 = [p(0)(1 - p(0))]^{1/2}$. Comparison between Eq. 3.73 and Eq. 3.79 shows that the WKB solution matches up with the *first* term in (3.73); the second (constant) term is important only near the boundary. And by equating coefficients in (3.73) and (3.79) we get

$$C_3 = [K(\alpha)/K(0)]e^{-NW(\alpha)}[p(\alpha)/(1 - p(\alpha))]^{-\lfloor N\alpha\rfloor}. \qquad (3.80)$$

The approximations (3.72) and (3.73) may now be substituted into the denominator and numerator of (3.71) respectively. (The numerator comprises only one term, since $1 - \sum_{z\in Q} T_{yz}$ is nonzero only if $y = \lfloor N\alpha\rfloor - 1$, in which case it equals $p(\alpha)$.) A bit of algebra yields a simple expression for $1 - \lambda_1$, and using the fact that $\mathbf{E}\tau \sim (1 - \lambda)^{-1}$ in the large-N limit we get the following. (We assume that $N\alpha$ is an integer; otherwise trivial changes are necessary.)

THEOREM 3.23 *In the one-process, one-resource (i.e., $q = r = 1$) memory model, when memory units are allocated and deallocated singly the mean time until exhaustion has asymptotics*

$$\mathbf{E}\tau \sim \sqrt{\frac{p(0)p(\alpha)(1 - p(\alpha))}{1 - p(0)}}(1 - 2p(\alpha))^{-1}\left(\frac{1}{1 - 2p(0)} + \frac{1}{p_0}\right)\exp{(NW(\alpha))}$$
$$(3.81)$$

as $N \to \infty$. The exponential growth rate $W(\alpha)$ may be computed from (3.75).

So the growth rate is $W(\alpha)$, but the pre-exponential factor is more complicated, and depends on the parameter p_0. (If the constraint of single allocations and deallocations is dropped, the exponential growth rate generalizes to $\int_0^\alpha p^*(z)\,dz$.) The formula (3.81), in slightly different notation, appears as Theorem 8.1 of Maier [Maier, 1991b]. But the present derivation is much cleaner.

7.4.3   ANALYSIS OF THE GENERAL MODEL

By building on the results of the preceding section, we can analyse the shared memory exhaustion model with an arbitrary number of processes q and resources (*i.e.*, memory types) r. We shall see that the technique of Section 7.4.1 yields the complete asymptotics of $E\tau$. It can in fact be viewed as an extension of the Wentzell-Freidlin approach used by Maier [Maier, 1991b], which yielded only the exponential growth rate.

First consider the case of a single resource. The state space and normalized state space are polyhedra; respectively

$$Q = \{\mathbf{y} \in \mathcal{N}^q : 0 \le y_i \le \lfloor N\alpha'' \rfloor, \sum_{i=1}^{q} y_i \le \lfloor N\alpha' \rfloor \}, \quad (3.82)$$

$$\mathcal{Q} = \{\mathbf{x} \in \mathcal{R}^q : 0 \le x_i \le \alpha'', \sum_{i=1}^{q} x_i \le \alpha' \}. \quad (3.83)$$

(We assume $q\alpha'' > \alpha'$, so the exhaustion face of $\mathcal{Q}$ is nonempty.) The system state evolves as follows. A process P_i is selected equiprobably from the set $\{P_1, \ldots, P_q\}$, and the corresponding y_i is incremented by ξ_i. As in the $q = 1$ model, the distribution of ξ_i may depend on $x_i \stackrel{\text{def}}{=} y_i/N$.

We first consider the case when ξ_i has density given by (3.70), with no dependence at all on x_i; $p < \frac{1}{2}$, so the state $\mathbf{0}$ is stable. This is the case of single allocations and deallocations, with their respective probabilities taken independent of the number of resource units currently allocated. We must modify the transition probabilities on the reflecting faces of Q; in particular, the faces where one or more of the y_i equals zero. If the process P_i has no quantity of the resource allocated, we take $\xi_i = \Delta y_i$ equal to $+1$ with probability p_0, and to 0 with probability $1 - p_0$. Here $p_0 \in (0, 1]$ is a 'hardness' parameter, as in the $q = 1$ model: it specifies the probability that complete deallocations are immediately followed by allocations.

This choice of reflecting boundary conditions makes the random processes $y_i(s)$ as independent as possible of one another near the equilibrium state $\mathbf{0}$. The quasi-stationary density ρ in Region 1, approximated as the solution of $\rho\mathbf{T} = \rho$, will be

$$\rho_{\mathbf{y}} = (p/p_0)^{|\mathbf{y}|_0} [p/(1-p)]^{\sum_{i=1}^{q} y_i} \quad (3.84)$$

which generalizes the constant-p case of (3.72). Here $|\mathbf{y}|_0$ is the number of zero components of the vector $\mathbf{y} = (y_1, \ldots, y_q)$, and the parameter p_0 affects $\rho_{\mathbf{y}}$ only if $|\mathbf{y}|_0 > 0$.

In the intermediate Region 2, which is the interior of $\mathcal{Q}$, a WKB solution is easily constructed. First note that if $p(x)$ is independent of x, the $q = 1$ WKB solution of (3.74), (3.75) and (3.76) specializes to

$$\rho_y = C_2'[p/(1-p)]^y \qquad (3.85)$$

for C_2' some constant. (This solution clearly satisfies (3.77) in the interior of Q.)

$$\rho_y = C_2'[p/(1-p)]^{\sum_{i=1}^q y_i} \qquad (3.86)$$

is the generalization to the $q > 1$ case.

As in the $q = 1$ case, on account of absorption of probability on the exhaustion face we set $\rho_y = 0$ for all y satisfying $\sum_{i=1}^q y_i = \lfloor N\alpha' \rfloor$. Moreover we modify the transition matrix elements T_{yz} of the Markov chain when y is one unit away from the boundary. If $\sum_{i=1}^q y_i = \lfloor N\alpha' \rfloor - 1$ we take:

$$T_{yz} = \begin{cases} q^{-1}(1-p), & \text{if } z_i = y_i - 1 \text{ for exactly one } i \\ 0, & \text{if } z_i = y_i + 1 \text{ for exactly one } i \end{cases}$$

where the zero probability assigned to each of the q possible allocations would normally be $q^{-1}p$. This alteration makes the transition matrix substochastic. With these choices, the approximation

$$\rho_y = C_3 \left([p/(1-p)]^{\sum_{i=1}^q y_i} - [p/(1-p)]^{\lfloor N\alpha' \rfloor} \right) \qquad (3.87)$$

will satisfy $\rho T = \rho$ in Region 3.

The WKB solution (3.86) interpolates between the approximations (3.84) and (3.87) of Regions 1 and 3. Comparison between (3.84) and (3.86) gives $C_2' = 1$. Comparison between (3.86) and the first term of (3.87) yields $C_3 = 1$. The second term is important, in a relative sense, only near the boundary.

Now that $C_3 = 1$ is known, the approximations (3.84) and (3.87) may be substituted into the denominator and numerator of (3.71). The summands in the numerator are nonzero only if y is one unit away from the boundary, in which case the factor $1 - \sum_{z \in Q} T_{yz}$ equals p. The number of such nonzero summands (they are all equal to each other) is to leading order the number of states on the exhaustion face of Q. This is N^{q-1} times a combinatorial factor, essentially the area of a truncated simplex in $\mathcal{R}^q$. Let us denote this factor $A(\alpha', \alpha'', q)$. After simplifying the expression (3.71), and using the fact that $E\tau \sim (1 - \lambda_1)^{-1}$, we get the following. (We assume $N\alpha'$ is an integer.)

THEOREM 3.24 *In the one-resource (i.e., $r = 1$, q arbitrary) shared memory model, when memory units are allocated and deallocated singly,* *with probabilities that are independent of current allocations, the mean time until exhaustion has asymptotics*

$$\mathbf{E}\tau \sim (1 - 2p)^{-1}p^q \left(\frac{1}{1 - 2p} + \frac{1}{p_0} \right)^q A(\alpha', \alpha'', q)^{-1} N^{-(q-1)}((1-p)/p)^{N\alpha'}$$

(3.88)

as $N \to \infty$. The quantity $A(\alpha', \alpha'', q)$ is computed as stated above.

The $q = 2$ case of this theorem was previously obtained as Theorem 2.13 of Louchard and Schott [Louchard and Schott, 1991] (in the special case when α'' is effectively infinite, and $p_0 = p$). But the $q > 2$ case is new. We see that if $q > 1$ the asymptotics of $\mathbf{E}\tau$ will necessarily include a power law factor, as well as the leading (dominant) exponential growth.

It follows that in the $N \to \infty$ limit, the system state at exhaustion time will be uniformly distributed across the exhaustion face of Q. This is because the rate of absorption of probability is, by the equality of the nonzero summands in the numerator in (3.71), independent of position on the face. So when p is constant, the limiting uniform distribution originally discovered by Flajolet [Flajolet, 1986] occurs irrespective of the number of processes.

Theorem 3.24 may be generalized in several directions; for example, to the case of batch allocations and deallocations of memory units. If the q processes can allocate and deallocate units in multiples rather than singly, $\xi_i = \Delta y_i$ will take on values other than ± 1, and formula (3.88) must be modified. The asymptotics will however be qualitatively similar: an exponential growth of $\mathbf{E}\tau$ with N, and a pre-exponential factor proportional to $N^{-(q-1)}$. The details may appear elsewhere.

It is more interesting to consider the generalization of Theorem 3.24 to the case when the probability of an allocation by a process P_i is allowed to depend on the currently allocated resource fraction $x_i = y_i/N$. In this case the transition probabilities T_{xy} will vary over the interior of Q. A WKB approximation $K(\mathbf{x}) \exp(-NW(\mathbf{x}))$ to the quasi-stationary density in Region 2 may still be constructed, but if $q > 1$ its behavior will be quite different from that of the WKB solution (3.86). We discuss this matter briefly.

The function $W(\mathbf{x})$ necessarily satisfies a so-called eikonal equation, a nonlinear partial differential equation on Q. (See Ref. [Knessl et al., 1985]; also Section 3 of Ref. [Naeh et al., 1990] for the continuous-time case.) This equation is the same as the Hamilton-Jacobi equation solved by Maier [Maier, 1991b], the solution of which is the 'quasi-potential' or 'classical action function' of Wentzell-Freidlin theory. Maier showed (in the $q = 2$ colliding stacks case, but the proof generalizes) that if the density of each ξ_i is given by (3.70), but with p taken to be a *decreasing* function of the allocation fraction x_i, then the behavior of $W(\mathbf{x})$ will be as follows. It will not depend on $\sum_{i=1}^q x_i$ alone, so it will not be constant over the exhaustion face. It will attain a quadratic minimum at the center of the face, where $\mathbf{x} = q^{-1}\alpha'(1, 1, \dots)$. The WKB solution, restricted to the exhaustion face, will accordingly display a *gaussian falloff* around this point; the standard deviation of this gaussian will be of order $N^{1/2}$ units.

The approximation in Region 3, with which the WKB solution must match, will exhibit a similar transverse falloff. So the sum in the numerator of (3.71), used in computing $1 - \lambda_1$, will be of magnitude $N^{(q-1)/2}$ rather than of magnitude N^{q-1}. The following is a consequence.

THEOREM 3.25 *In the one-resource (i.e., $r = 1$, q arbitrary) shared memory model, when memory units are allocated and deallocated singly, with the probability of an allocation by each process assumed to be a decreasing function of the fraction of memory units allocated to it (the 'increasingly contractive' case), the mean time until exhaustion has $N \to \infty$ asymptotics*

$$\mathbf{E}\tau \sim C(\alpha', \alpha'', q)N^{-(q-1)/2}((1-p)/p)^{N\alpha'}. \qquad (3.89)$$

It is not a trivial matter to compute the constant $C(\alpha', \alpha'', q)$ in (3.89) when $q > 1$. (This requires solving the 'transport equation' [Knessl et al., 1985] satisfied by the pre-exponential factor $K(\mathbf{x})$ in the WKB solution.) However, it follows from our derivation that the system state at exhaustion time will be within $O(N^{1/2})$ units of the state $q^{-1}\alpha' N(1, 1, \dots)$. To within this accuracy, the processes $P_1, \dots, P_q$ will have *equal* quantities of the resource R_1 on hand when exhaustion occurs. We emphasize that this is special to the 'increasingly contractive' case, in which $p(x)$ is a decreasing function of x.

We state without proof the generalization of Theorems 3.24 and 3.25 to the case of an arbitrary number of resources r.

THEOREM 3.26 *In the general shared memory model with r types of memory, when memory units are allocated and deallocated singly, the $N \to \infty$ asymptotics of the mean time until exhaustion will be given by formula (3.88) (in the case when allocation probabilities are independent of the number of resource units allocated) or by formula (3.89) (in the case when they are a decreasing function of the fraction allocated). In both cases the formula for $\mathbf{E}\tau$ must be multiplied by a factor r^{-1}.*

We can expect that asymptotic techniques similar to those of this section will prove useful in analysing the behavior of stochastically modelled on-line algorithms whenever (1) the algorithm state space is naturally viewed as a subset of some finite-dimensional Euclidean space, and (2) there is some scaling parameter N, governing the size of this set, which is being taken to infinity. This includes the analysis of buddy systems and related storage allocators. Many years ago Purdom and Stigler [P. W. Purdom and Stigler, 1970] began the stochastic analysis of binary buddy systems. The approach of this section may well permit the computation, in the case of an arbitrary number of block sizes, of the asymptotics of the expected time until memory exhaustion.

8.   REGULAR APPROXIMATION OF SHUFFLE PRODUCTS OF CONTEXT-FREE LANGUAGES

The exposition in this section is based on [Maier and Schott, 1993b].

8.1   INTRODUCTION

The concept of a *structure generating function* of a formal language is well known [Flajolet, 1987, Salomaa and Soittola, 1978]. If $L \subseteq \Sigma^*$ is a language over $\Sigma = \{a_1, \ldots, a_k\}$, the simplest such generating function is a formal series in a single indeterminate z,

$$L(z) \overset{\text{def}}{=} \sum_{w \in L} z^{|w|}. \tag{3.90}$$

Here the coefficient L_n of z^n in $L(z)$ simply counts the number of words of length n in L. A *multivariate* generating function would be a formal series in k commuting indeterminates $z_1, \ldots, z_k$, *i.e.*,

$$L(z_1, \ldots, z_k) = \sum_{w \in L} z_1^{|w|a_1} \cdots z_k^{|w|a_k}. \tag{3.91}$$

The coefficient $L_{n_1, \ldots, n_k}$ of $z^{n_1} \cdots z^{n_k}$ is the number of words with n_1 a_1's, n_2 a_2's, *etc.* Abstractly, these generating functions are $\mathcal{N}$-valued functions on the free *abelian* monoids generated by z and by $z_1, \ldots, z_k$. There are close ties between the theory of structure generating functions and the theory of functions of a complex variable. If L is *regular*, the formal series for $L(z)$ has positive radius of convergence, and converges to a rational function of z. The class of rational functions which can arise in this way was determined by Berstel and Soittola (see the references in [Maier and O'Cinneide, 1992, Salomaa and Soittola, 1978]). If L is a context-free language (CFL) which is not inherently ambiguous, then the series for $L(z)$ converges to an algebraic function. Flajolet [Flajolet, 1987] has used this to prove that several context-free languages are inherently ambiguous.

Here we employ generating functions, in general multivariate, to quantify the degree of closeness between certain deterministic CFL's and their *regular approximants*: the regular languages one gets by restricting the push-down automata recognizing them to have maximum storage size $N < \infty$. We are especially interested in quantifying the rate of convergence of the regular approximants as $N \to \infty$. For this we use an asymptotic technique developed by Naeh *et al.* [Naeh et al., 1990] in the context of the exit problem for Markov chains.

A full analysis of this convergence would presumably require topologizing or metrizing the spaces of rational and algebraic functions which can arise as generating functions. No natural way of doing this is known: the rational functions are normally meromorphic rather than entire, and the algebraic functions are not even defined, strictly speaking, on $\mathcal{C}$ or $\mathcal{C}^k$. That is because the series (3.90) and (3.91) converge only in a neighborhood of 0, and their analytic continuation is to a more general Riemann surface.

We shall nonetheless quantify the rate of convergence by studying one key property of the generating functions: the location in $\mathcal{C}$ (or $\mathcal{C}^k$) of their nearest singularity to 0, or equivalently their radii of convergence. These radii have a straightforward interpretation in terms of the language L. The univariate series (3.90) has radius of convergence

$$r = \left(\limsup_{n \to \infty} L_n^{1/n}\right)^{-1} \tag{3.92}$$

so r^{-1} measures the (asymptotic) *exponential growth rate* at which L_n grows with n. The multivariate case is similar: if $z_i = z\alpha_i$ for $\{\alpha_i\}_{i=1}^k$ a fixed positive sequence, we may study the convergence of (3.91) as a function of z. That is, we may compute the least $z > 0$ for which the series for $L(z\alpha_1, \ldots, z\alpha_k)$ fails to converge. The coefficient of z^n in $L(z\alpha_1, \ldots, z\alpha_k)$ is

$$L_n(\alpha_1, \ldots, \alpha_k) \stackrel{\text{def}}{=} \sum_{|w|=n} \alpha_1^{|w|_{a_1}} \cdots \alpha_k^{|w|_{a_k}} \tag{3.93}$$

rather than L_n, and the reciprocal of the radius of convergence measures the (asymptotic) exponential growth rate of these coefficients. Using a multivariate rather than a univariate generating function allows us to weigh the letters in Σ differently, as in (3.93); if $\alpha_i \equiv 1$ the multivariate case reduces to the univariate.

The following makes it clear that these radii of convergence have an interpretation in terms of the eigenvalues, or in general the spectra or spectral radius, of certain transition matrices. If L is recognized by a single-stack deterministic push-down automaton (DPDA), we denote the

DPDA by $(Q, X, i, F, \Sigma, \delta)$. Here X is the stack alphabet, Q the auxiliary finite state space, i the initial state and F the set of final states. (The automaton is assumed to recognize by empty stack.) The full DPDA state space is $Q \times X^*$, and the transition map $\delta \colon Q \times X \times \Sigma \to Q \times X^*$ induces a map from $(Q \times X^*) \times \Sigma$ to $Q \times X^*$ describing how the DPDA state alters when a new letter is seen. Denote by $T_{a_i} \colon Q \times X^* \to Q \times X^*$ the transition induced by the letter a_i. T_{a_i} may be viewed as an infinite matrix over $\{0, 1\}$, indexed by $Q \times X^*$, with exactly one 1 per row since the automaton is by assumption deterministic. It is clear, given the recognition by empty stack, that

$$L(z_1, \ldots, z_k) = \mathbf{1}_{(1,\epsilon)}^t (\mathbf{I} - z_1 \mathbf{T}_{a_1} - \cdots - z_k \mathbf{T}_{a_k})^{-1} \mathbf{1}_{(F,\epsilon)}. \qquad (3.94)$$

Here $(\mathbf{I} - z_1 \mathbf{T}_{a_1} - \cdots - z_k \mathbf{T}_{a_k})^{-1}$ is a matrix indexed by $Q \times X^*$, and $\mathbf{1}_{(F,\epsilon)}$ is a column vector indexed by $Q \times X^*$ which is equal to 1 in the states (f, ϵ), for $f \in F$, and 0 otherwise. $\mathbf{1}_{(1,\epsilon)}^t$ is a row vector equal to 1 in state (i, ϵ), and 0 otherwise.

Since we are taking $z_i = z\alpha_i$ for all i, we write

$$L(z_1, \ldots, z_k) = \mathbf{1}_{(1,\epsilon)}^t (\mathbf{I} - z|\Sigma|\mathbf{T})^{-1} \mathbf{1}_{(F,\epsilon)} \qquad (3.95)$$

where

$$\mathbf{T} \overset{\text{def}}{=} |\Sigma|^{-1} \sum_i \alpha_i \mathbf{T}_{a_i}. \qquad (3.96)$$

The representation (3.95) makes it clear that the radius of convergence of the Maclaurin series in z for $L(z_1, \ldots, z_k)$ is related to the spectral radius of $\mathbf{T}$. In fact we expect that the radius of convergence will be $|\Sigma|^{-1} \|\mathbf{T}\|^{-1}$, for $\|\mathbf{T}\|$ the spectral radius. If $\mathbf{T}$ were a finite matrix the spectral radius would be $\max_j |\lambda_j(\mathbf{T})|$, the maximum of the moduli of its eigenvalues. But $\mathbf{T}$ here is an infinite matrix, so it is in general harder to compute.

Note however that if $\sum_i \alpha_i = |\Sigma|$ (an innocuous normalization constraint which we assume henceforth) then $\mathbf{T}$ will be a stochastic matrix: the sum of every row will be unity. It follows by the Perron-Frobenius theorem [Minc, 1988] that $\|\mathbf{T}\| = 1$. So the radius of convergence equals $|\Sigma|^{-1}$. And the asymptotic exponential growth rate of the coefficients $L_n(\alpha_1, \ldots, \alpha_k)$ will be $|\Sigma|$, irrespective of the choice of $(\alpha_1, \ldots, \alpha_k)$. In particular, to leading order $L_n \sim |\Sigma|^n$. This is ultimately a consequence of determinism.

Restricting the stack to have height at most N will, however, yield a *non*-deterministic finite automaton. Finite square submatrices of $\mathbf{T}$ are substochastic, not stochastic: not all their rows will have unit sum.

This indicates that the singularity structure of $L(z)$ and $L(z_1, \ldots, z_k)$ (at least, the singularities closest to 0, which are dominant and also the most accessible) will differ between L and its rational approximants.

The above-mentioned technique of Naeh *et al.* is in essence a technique for computing the asymptotics of the largest eigenvalue of certain substochastic (but nearly stochastic) matrices, the transition matrices of Markov chains with absorption. So it finds an application here. In the next section we introduce the languages L we shall analyse, and draw the necessary parallels with the theory of Markov chains. In Section 7.4.1 we explain the technique, and in Section 8.3 we apply it.

8.2   DYCK LANGUAGES AND SHUFFLE PRODUCTS

In general the DPDA state space $Q \times X^*$ of the previous section is quite large, and it is difficult to study Markov chains on it. Following Flajolet [Flajolet, 1987], we shall simplify matters by considering only certain *multi-counter* deterministic CFL's: languages recognized by multistack automata in which each stack has only a single stack symbol. In effect the multistack state space will be $\mathcal{N}^d$, and will have a Euclidean structure.

Our languages will be derived from the usual Dyck (*i.e.*, parenthesis) language D over $\Sigma = \{a_1, a_2\}$, generated by the production

$$D \to a_1 D a_2 D + \epsilon. \tag{3.97}$$

The corresponding univariate generating function is

$$D(z) = \sum_{n=0}^{\infty} D_n z^n = \frac{1 - \sqrt{1 - 4z^2}}{2z^2}. \tag{3.98}$$

$D(z)$ admits $z = \frac{1}{2}$ and $z = -\frac{1}{2}$ as singularities. Expanding $D(z)$ we get the Catalan numbers

$$D_{2m} = \frac{1}{m+1} \binom{2m}{m} \sim \frac{2^{2m}}{\sqrt{\pi m^3}}, \tag{3.99}$$

i.e., $D_{2m} \sim 2^{2m}$ to leading order. D is a one-counter CFL: it is recognized by a PDA with a one-state Q and a one-letter stack alphabet X. The PDA state space is X^*, *i.e.*, $\mathcal{N}$; the left parenthesis map $\mathbf{T}_{a_1}$ increases the stack state by 1, and the map $\mathbf{T}_{a_1}$ decreases it.

This PDA is not deterministic (a right parenthesis cannot be handled when the stack is empty). So instead of D we consider the modified Dyck language

$$E \stackrel{\text{def}}{=} a_2{}^*(Da_2{}^*)^* \tag{3.100}$$

which allows an arbitrary number of right parentheses when the stack is empty. The corresponding PDA is deterministic, and E has univariate generating function

$$E(z) = \left[1 - z - D(z)\right]^{-1} \qquad (3.101)$$

which has the same singularities as $D(z)$. It is easily checked that $E_n \sim 2^n$ to leading order, *i.e.*, $E_n \sim |\Sigma|^n$. (This is guaranteed by the determinism.) E is a more interesting language than D from the generating function point of view. Since $|w|_{a_1} = |w|_{a_2}$ for all $w \in D$, the bivariate generating function of D is trivially expressible in terms of the univariate generating function. This is not the case with E.

If the automaton state space is identified with $\mathcal{N}$, for any $\{\alpha_i\}$ the transition matrix $\mathbf{T} = (T_{xy})$ of E defined by (3.96) will be given by

$$\begin{aligned} T_{xy} &= \tfrac{1}{2}(\alpha_1 \delta_{y,x+1} + \alpha_2 \delta_{y,x-1}), & x > 0 \qquad (3.102) \\ T_{xy} &= \tfrac{1}{2}(\alpha_1 \delta_{y,x+1} + \alpha_2 \delta_{y,x}), & x = 0. \end{aligned}$$

For any fixed $N \in \mathcal{N}$, the N'th rational approximant to E (which we write as $E^{[1,N]}$, in a notation to be explained below) by definition has state space $\{0, \ldots, N\}$, and the elements of its transition matrix $\mathbf{T}^{[1,N]}$ are given by (3.102) for all $x, y \in \{0, \ldots, N\}$. As noted, the square submatrices $\mathbf{T}^{[1,N]}$ of $\mathbf{T}$ are substochastic. But like $\mathbf{T}$ they are tridiagonal (*i.e.*, Jacobi) matrices. It is not difficult to diagonalize them, and to compute their spectral radii explicitly as a function of N and (α_1, α_2).

We are interested in less trivial automata, whose associated transition matrices cannot be explicitly diagonalized but can be studied asymptotically. For this reason, following Flajolet [Flajolet, 1987] we consider multicounter languages obtained by *shuffle products*. Recall that the shuffle product $L^{(1)} \amalg L^{(2)}$ of languages $L^{(1)}$ and $L^{(2)}$ (assumed to be over disjoint alphabets) is the set of all words obtained by interleaving words from $L^{(1)}$ and $L^{(2)}$. On the level of generating functions, we have

$$\left(L^{(1)} \amalg L^{(2)}\right)_n = \sum_{p+q=n} \binom{n}{p} L^{(1)}{}_p L^{(2)}{}_q. \qquad (3.103)$$

The d-fold shuffle product $D^{[d]} \overset{\text{def}}{=} \amalg^d D$ of the usual Dyck language is a parenthesis language over an alphabet $\Sigma = \{a_1, \ldots, a_{2d}\}$ containing d sorts of parenthesis pairs, generated by the production

$$D^{[d]} \to a_1 D^{[d]} a_2 D^{[d]} + \cdots + a_{2d-1} D^{[d]} a_{2d} D^{[d]} + \epsilon. \qquad (3.104)$$

The shuffle product $E^{[d]} \overset{\text{def}}{=} \amalg^d E$ has a similar interpretation.

$E^{[d]}$, like $D^{[d]}$, is a d-counter language: it is recognized by an automaton whose state space is $\mathcal{N}^d$. The stochastic transition matrix $\mathbf{T}^{[d]} = (T^{[d]}_{\mathbf{xy}})$, for any choice of $\{\alpha_i\}_{i=1}^{2d}$, is defined by a formula which generalizes (3.102):

$$T^{[d]}_{\mathbf{xy}} = |2d|^{-1} \sum_{j=0}^{d-1} (\alpha_{2j+1}\delta_{\mathbf{y},\mathbf{x}+\mathbf{e}(2j)} + \alpha_{2j+2}\delta_{\mathbf{y},\mathbf{x}-\mathbf{e}(2j+1)}). \qquad (3.105)$$

Here $\mathbf{e}(j)$ denotes the j'th unit vector $(0,\dots,1,\dots,0)$, and $\mathbf{x} - \mathbf{e}(i)$ is understood to equal $\mathbf{x}$ whenever $x_i = 0$. We shall restrict ourselves to $\{\alpha_i\}$ with the property that α_i depends only on $i \bmod 2$, so that all left parentheses are weighted equally, as are all right parentheses.

As Flajolet [Flajolet, 1987] discusses, the languages $E^{[d]}$ and $D^{[d]}$ occur naturally in the context of *storage allocation algorithms*. Each word $w \in D^{[d]}$ can be viewed as a sequence of allocations and deallocations of d types of resource. Allocation of each resource follows a stack-like (LIFO) discipline, but interleaving is allowed. $E^{[d]}$ has moreover a very natural Markov chain interpretation: its transition matrix $\mathbf{T}^{[d]}$ is stochastic, and for any $l \in \mathcal{N}$, the matrix $\mathbf{T}^l$ specifies the evolution of a stochastically modelled allocation system after l time units.

It was in this context that rational approximants were first introduced. The natural way of approximating the d-counter language $E^{[d]}$ by a rational language $E^{[d,N]}$ is to require that the *total* stack space used in recognizing words $w \in E^{[d,N]}$ be less than or equal to N. (This is a constraint on the total space utilization of the allocated objects.) That is, the state space for $E^{[d,N]}$ will be

$$Q' \overset{\text{def}}{=} \{\, \mathbf{x} \in \mathcal{N}^d \mid 0 \leq \sum_{i=1}^{d} x_i \leq N \,\}. \qquad (3.106)$$

If $d > 1$, the substochastic matrix $\mathbf{T}^{[d,N]}$ obtained by restricting the transition matrix $\mathbf{T}^{[d]}$ of (3.105) to this polytope unfortunately cannot be diagonalized explicitly. This is in part due to the fact that $E^{[d,N]} \neq \amalg^d E^{[1,N]}$, *i.e.*, the taking of rational approximations does not commute with the shuffle product operation.

Results have nonetheless been obtained by Flajolet [Flajolet, 1986], Louchard and Schott [Louchard and Schott, 1991], and Maier [Maier, 1991b] on the large-N asymptotics of $1 - \|\mathbf{T}^{[N,d]}\|$, in the $d = 2$ ("colliding stacks") case. (The quantity $1 - \|\mathbf{T}^{[N,d]}\|$, which is a function of (α_1, α_2), measures the distance between the generating functions of $E^{[d]}$ and $E^{[d,N]}$.) As is sketched in the next section, when N is large

$1 - \|\mathbf{T}^{[N,d]}\|$ is essentially an *absorption rate* for the Markov chain determined by $\mathbf{T}^{[N,d]}$: the rate at which probability is absorbed at the boundary $\{\mathbf{x} \in \mathcal{N}^d \mid \sum_{i=1}^{d} x_i = N\}$ of the state space Q' due to substochasticity. The qualitative behavior of the Markov chain depends strongly on the relative sizes of α_1 and α_2 (the left and right parenthesis weightings, respectively). If $\alpha_1 > \alpha_2$, one expects the absorption will take place on the whole in linear time, i.e., $1 - \|\mathbf{T}^{[N,d]}\| = \Theta(N^{-1})$, $N \to \infty$. If $\alpha_1 = \alpha_2$ the chain is in a sense *diffusive*: the combinatoric results of Flajolet imply that $1 - \|\mathbf{T}^{[N,2]}\| = \Theta(N^{-2})$. The case $\alpha_1 < \alpha_2$ is the hardest and the most interesting. Here the Markov chain defined by (3.105) is recurrent, and the state $(0, \ldots, 0) \in \mathcal{N}^d$ is an equilibrium state. Absorption on the boundary is correspondingly difficult, occurring only if left parentheses (which are suppressed) predominate over right parentheses. The absorption rate is accordingly small. Flajolet showed, in our notation, that $1 - \|\mathbf{T}^{[N,2]}\| = \Theta((\alpha_2/\alpha_1)^{-N})$.

Louchard and Schott investigated the $\alpha_1 < \alpha_2$ case of the $d = 2$ problem further, and by probabilistic techniques obtained the constant factor in $1 - \|\mathbf{T}^{[N,2]}\|$ as well as the exponential decay rate as $N \to \infty$. The $d > 2$ problem is apparently very difficult to treat either by combinatoric or standard probabilistic methods. In Section 8.3 we show however that when d is arbitrary, in the $\alpha_1 > \alpha_2$ case

$$1 - \|\mathbf{T}^{[N,d]}\| \sim c_d N^{-d+1} (\alpha_1/\alpha_2)^{-N} \qquad (3.107)$$

for a (α_1, α_2)-dependent constant c_d which can be computed explicitly. For this we use the technique of Naeh *et al.* for computing the formal asymptotics of Markov chain absorption rates, or first passage times. The result (3.107) solves an open problem on rate of convergence of the rational approximants $E^{[d,N]}$ to $E^{[d]}$ as $N \to \infty$.

8.3   THE ANALYSIS

The Markov chain on the state space Q' of (3.106), with (α_1, α_2)-dependent transition matrix $\mathbf{T}^{[d,N]}$ given by the restriction of (3.105) to Q', fits naturally into the framework of the last section if $\alpha_1 < \alpha_2$. (If $\alpha_1 < \alpha_2$ then $(0, \ldots, 0)$ is stable as desired.) The N-dependent state space Q' can be viewed as arising from the normalized state space

$$Q = \{\mathbf{z} \in \mathcal{R}^d \mid 0 \le \sum_{i=1}^{d} z_i \le 1\} \qquad (3.108)$$

by scaling. In the interior of Q' the distribution of the increment random variable ξ is actually independent of state; clearly

$$P(\xi = +\mathbf{e}(j)) = \alpha_1/2d, \quad P(\xi = -\mathbf{e}(j)) = \alpha_2/2d \qquad (3.109)$$

for all $1 \leq j \leq d$. Recall that $\alpha_1 + \alpha_2 = 2$ by assumption.

Of course due to the 'reflecting' empty-stack boundary condition, the distribution of ξ differs if one or more components of the state $\mathbf{x}$ equals zero. Also, if $\mathbf{x}$ is on the absorbing boundary then $\sum_{\mathbf{y}} \mathbf{T}_{\mathbf{xy}}^{[d,N]} < 1$.

Although the matrices $\mathbf{T}^{[1,N]}$ can be diagonalized exactly, we work out the large-N asymptotics of $1 - \lambda_1$ when $d = 1$ to illustrate the technique. If $d = 1$ the random increment ξ takes ± 1 values, and $P(\xi = +1) = \alpha_1/2$. We approximate the quasi-stationary density ρ as the solution of $\rho \mathbf{T}^{[1,N]} = \rho$; this is the same as assuming, for the moment, that $\lambda_1 \approx 1$. In Region 1, where $x = o(N)$, it is clear that

$$\rho_x = (\alpha_2/\alpha_1)^{-x} \tag{3.110}$$

satisfies $\rho \mathbf{T}^{[1,N]} = \rho$. The behavior of ρ_x in Region 3, where $x = N - o(N)$, is similar. We require $\rho_{N+1} = 0$ since probability, once absorbed on the boundary, never returns. The approximation

$$\rho_x = C[(\alpha_2/\alpha_1)^{-x} - (\alpha_2/\alpha_1)^{-N-1}] \tag{3.111}$$

will satisfy $\rho \mathbf{T}^{[1,N]} = \rho$ in Region 3. C must be chosen to ensure consistency between (3.110) and (3.111), but clearly $C = 1$ is the correct choice. A study of the behavior of ρ_x in the intermediate Region 2 shows that the second term in (3.111) has no counterpart in (3.110) because its continuation to Region 1 is exponentially small compared to the first term, and can be neglected.

The approximations (3.110) and (3.111) may be substituted into the denominator and numerator of (3.71) respectively. (The numerator comprises only a single term, since $1 - \sum_y T_{xy}$ is nonzero only if $x = N$, in which case it equals $(\alpha_1/2)\rho_N$.) We get

$$1 - ||\mathbf{T}^{[1,N]}|| = 1 - \lambda_1 \sim (\alpha_1/2)(\alpha_2/\alpha_1)^{-N}. \tag{3.112}$$

But the radius of convergence of the series for $L(z\alpha_1, z\alpha_2)$ equals $|\Sigma|^{-1}||\mathbf{T}||^{-1}$. We conclude that when $\alpha_2 > \alpha_1 > 0$, the closest singularity of the generating function $E^{[1,N]}(z_1, z_2)$ to $(0,0)$ in $\mathcal{C}^2$ along the line

$$(z_1, z_2) = z(\alpha_1, \alpha_2), \; z > 0$$

occurs when

$$z \sim \tfrac{1}{2}[1 + (\alpha_1/2)(\alpha_2/\alpha_1)^{-N}]. \tag{3.113}$$

This singularity is of course a pole, since $E^{[1,N]}(z_1, z_2)$ is a rational function of z. As $N \to \infty$ the location of this pole converges to $z = \tfrac{1}{2}$, which is the location of the dominant singularity of the *algebraic* generating function $E(\alpha_1 z, \alpha_2 z)$, when viewed as a function of z. We previously saw that $z = \frac{1}{2}$ is a singularity of the univariate generating function, *i.e.*, is a singularity in the degenerate $\alpha_1 = \alpha_2$ case.

This approach extends immediately to the $d > 1$ languages. We may approximate the quasi-stationary density ρ on $Q' \subset \mathcal{N}^d$ as the solution of $\rho = \rho \mathbf{T}^{[d,N]}$. In Region 1, where $x_i = o(N)$ for all i, we take

$$\rho_{\mathbf{x}} = (\alpha_2/\alpha_1)^{-\sum_{i=1}^{d} x_i}. \tag{3.114}$$

In Region 3, *i.e.*, within a distance $o(N)$ of the absorbing boundary, the approximation

$$\rho_{\mathbf{x}} = (\alpha_2/\alpha_1)^{-\sum_{i=1}^{d} x_i} - (\alpha_2/\alpha_1)^{N+1} \tag{3.115}$$

is appropriate; the behavior of (3.115) is consistent with the absorbing boundary condition. As in the $d = 1$ case the first terms of these approximations are identical. It is not a trivial matter, however, to work out the large-N behavior of the function in the intermediate Region 2 whose constant limit as $\sum_i x_i/N \to 1$ appears as the second term in Region 3.

The approximations (3.114) and (3.115) may be substituted into (3.71) as before. Now every state on the absorbing boundary contributes to the numerator, but to leading order the contributions will be equal. The cardinality of the set of absorbing states therefore enters; to leading order it is $N^{-d+1} A_d$, where A_d is the area of the unit $d-1$-dimensional simplex in $\mathcal{R}^d$:

$$\{\mathbf{z} \in \mathcal{R}_+^d \mid \sum_{i=1}^{d} z_i = 1\}. \tag{3.116}$$

The summations in (3.71) are easily performed, and we get

$$1 - \|\mathbf{T}^{[d,N]}\| \sim (\alpha_1/2)[1 - (\alpha_1/\alpha_2)]^{-d+1} A_d N^{-d+1} (\alpha_2/\alpha_1)^{-N} \tag{3.117}$$

which extends (3.112). We summarize the consequences in the following theorem.

THEOREM 3.27 *When $\alpha_2 > \alpha_1 > 0$, the closest singularity of the generating function $E^{[d,N]}(z_1, \ldots, z_{2d})$ to $(0,0)$ in $\mathcal{C}^{2d}$ along the line*

$$(z_1, \ldots, z_{2d}) = z(\alpha_1, \alpha_2, \alpha_1, \alpha_2, \ldots), \ z > 0$$

occurs when

$$z \sim (2d)^{-1}\{1 + (\alpha_1/2)[1 - (\alpha_1/\alpha_2)]^{-d+1} A_d N^{-d+1} (\alpha_2/\alpha_1)^{-N}\} \tag{3.118}$$

as $N \to \infty$.

This theorem, with its explicit formula for the constant factor in the deviation of the singularity from $(2d)^{-1}$, extends the $d = 2$ results of Louchard and Schott [Louchard and Schott, 1991] and solves a problem that has been open since the work of Flajolet [Flajolet, 1986].

8.4   CONCLUSIONS

It is clear from the computations that when the state space of an automaton recognizing a context-free language has a natural Euclidean structure, it is comparatively easy to approximate the dominant singularities of the multivariate generating functions of the regular approximants to the language. Theorem 3.27 is an example of the power of formal asymptotic methods in this regard.

However when the state space has no natural Euclidean structure, the problem may be much harder. And even for the multi-counter languages considered here, it is not clear how to control the location of *subdominant* poles as $N \to \infty$. In the large-N limit these poles in some sense merge to form the branch cuts present in algebraic generating function of the context-free language. This process remains unexplored.

9.   DESIGN AND ANALYSIS OF RANDOMIZED ALGORITHMS
9.1   GENERALITIES AND BASIC PRINCIPLES

A randomized algorithm realizes random choices during its execution. A randomized algorithm solves a deterministic problem and, whatever the result of choices done, the algorithm terminates in finite time and outputs a deterministic solution for the input problem. Therefore only the execution of the algorithm involves a part of randomness not its result. Randomized algorithms are rather simple, easy to code and efficient in practice. Moreover, they do not require that the input data satisfy some probabilistic distribution but only that they are inserted in random order. For many applications, a randomized algorithm is the simplest algorithm available, or the fastest, or both. Randomized algorithms are one-line. Clarkson and Shor [Clarkson and Shor, 1989] have given a general framework for the design and analysis of randomized incremental constructions. It provides general bounds for their expected space and time complexities when averaging over all permutations of the input data.

The last decade has witnessed a tremendous groth in the area of randomized algorithms. We specially recommend the books of J.D. Boissonnat and M. Yvinec [Boissonnat and Yvinec, 1995], K. Mulmuley [Mulmuley, 1993] and R. Motwani and P. Raghavan [Motvani and Raghavan, 1995] which present the basic concepts in the design and analysis of random-ized algorithms and describe many geometric applications. Our goal is much less ambitious since we just present a few randomized algorithms: first, following [Devillers, 1996] and [Mulmuley, 1993]. we take another view of quicksort then we focus on classical computational geometry ex-amples (convex hull, Voronoi diagrams, arrangements of planar curves) taken from [Devillers, 1996].

9.2   ANOTHER VIEW OF QUICKSORT

Let $S = \{x_i\}_{i \in \{0,...,n\}}$ the set to be sorted. We express the problem in terms of objects, regions and conflicts. In this case the objects are real numbers. A region is the interval $[x_i, x_j]$ determined by two objects x_i and x_j. We say that a number conflicts $[x_i, x_j]$ if it belongs to $]x_i, x_j[$. The first method consists in computing the result for the k first objects and storing the conflicts between the already computed regions and the non yet inserted objects in a structure called conflict graph [Clarkson and Shor, 1989]. In our example, this means that the k first numbers x_i, $i \leq k$ have been sorted and conflicts are known. Now we look at the insertion of x_{k+1}. We know, from the conflict graph, the interval $[x_i, x_j]$ containing it, so this interval is no longer without conflict. We have to delete it and create two new intervals $[x_i, x_{k+1}]$ and $[x_{k+1}, x_j]$. Then the $k + 1$ first numbers are sorted and the conflict graph has to be updated: numbers which were conflicting other intervals than $[x_i, x_j]$ are not affected, and those conflicting $[x_i, x_j]$ are now in conflict with either $[x_i, x_{k+1}]$ or $[x_{k+1}, x_j]$. Remark that the conflict graph algorithm for sorting is therefore just another point of view quicksort.. This conflict graph method has the disadvantage to be static: all objects must be known in advance to initialize the conflict graph with an unique region conflicting all objects.

9.2.1   A SEMI-DYNAMIC APPROACH

We replace the conflict graph by another structure which, instead of storing the conflicts of non yet inserted objects, locates the regions in conflict with the new object. This approach yields a semi-dynamic algorithm, objects are not known in advance but only when they are inserted. The basic idea of the influence graph [Boissonnat et al., 1992] consists in remembering the history of the construction. When the in-sertion of a new object makes the conflicting regions disappear, they are not deleted but just marked inactive. The regions created are linked to existing regions in the influence graph in order to locate further conflicts. In the case of sorting, the influence graph is a binary tree whose nodes are intervals, the two sons of a node correspond to the splitting of that interval into two sub-intervals. When a new number x_{k+1} is inserted, it is located in the binary tree, the leaf containing it becomes an internal node, its interval $[x_i, x_j]$ is split into two with respect to the new inserted number. Thus for sorting, the influence graph is a binary search tree (without balancing schema).

The comparisons done in the static and in the semi-dynamic algorithms are exactly the same: if x_i and x_j, $i < j$ must be compared, they are compared during the insertion of x_i in the conflict graph and during the insertion of x_j in the influence graph.

Remark.

The static and semi-dynamic algorithms presented above are not randomized. They are incremental algorithms, updating the set of regions without conflict each time a new object is inserted.

Now, we will randomize the algorithm, that is introduce some randomness: objects are no longer inserted in an order determined by the user but in a random order. On the example of sorting, we analyze now the average cost of inserting the nth number x_n in the influence graph. Just recall that all the intervals without conflict existing at a given time during the construction remain in the graph. Just after the insertion of the kth number, only one interval $[x_i, x_j]$ (without conflict at stage k) is in conflict with x_n, then the randomized hypothesis yields that x_i is the kth number with probability $\frac{1}{k}$ and the same for x_j. This means that the region conflicting x_n at stage k was created at stage k with probability $\frac{2}{k}$. Thus the number of regions conflicting x_n is easily obtained by summing over their creation stage. The cost of insertion of a number in the influence graph is in $O(logn)$ and the cost of sorting the n numbers is therefore in $O(nlogn)$.

Similar techniques apply in general cases and more formally [Clarkson and Shor, 1989]:

THEOREM 3.28 *If $f_0(k)$ is the expected number of regions without conflict determined by a random sample of size k of a set S of n objects, then regions without conflicts determined by S can be computed in time $O(\sum_1^n f_0(\lfloor \frac{n}{j} \rfloor))$ and space $O(\sum_1^n \frac{f_0(\lfloor \frac{n}{j} \rfloor)}{j})$, using conflict graph or influence graph technique provided that the objects are inserted in a random order.*

COROLLARY 2 *If $f_0(k) = O(k)$ then the regions without conflict can be determined in $O(nlogn)$ expected time and $O(n)$ expected space.*

9.2.2   FULLY DYNAMIC ALGORITHM

By fully dynamic we mean that objects can be inserted into and removed from the structure. Several techniques have been used to solve this important but difficult problem. We focus below on the coin flipping method which involves Pug's skip list data structure [Pugh, 1990]. See also Papadakis et al [Papadakis et al., 1992] and Devroye [Devroye, 1992] for carefully analyses of this data structure.

Assume that $x_1 < x_2 < \ldots x_n$ are n sorted numbers. A sample of this set is constructed by tossing a coin for each number to decide if it belongs to the sample or not. Then the selected set is sampled again in the same manner, and so on until there is no number left. The different levels of the structure are linked together, and if another number must be located in the set, it is successively located at each level. The location at one level is used as starting point for the location at the level below.

The above description is static, but this data structure can be dynamically maintained: when a new number is inserted, it is first located at the lower level, inserted at this level and then a coin is flipped to decide if it has to be inserted above, and so on while the coin tossing is successful. When an object must be deleted, it is enough to delete it at each level where it appears. The update is also very easy. This data structure is a kind of tree with non fixed arity. The structure is balanced on average and has constant arity on average

9.3   APPLICATIONS IN COMPUTATIONAL GEOMETRY

9.3.1   CONVEX HULL

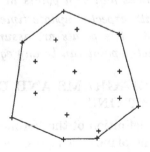

Figure 3.12.   Convex hull

Given a set S of n points, their convex hull is the smallest convex set that contains all of the n points (see Figure 3.12).

Semi-dynamic algorithm. To apply the general framework of the influence graph to planar convex hull we define objects as points, regions as half-planes defined by two points. We say that a point conflicts a half-plane if it belongs to it.

The convex hull problem reduces to the computation of non conflicting regions. The region defined by p and q has no conflict if and only if the segment pq is an edge of the convex hull. Here $f_0(k) = O(k)$ (the size of the convex hull is linear) and application of Theorem 3.28 yields:

THEOREM 3.29 *The influence graph maintains the convex hull of n points in the plane in expected time $O(\log n)$ per insertion, if the order of insertion is random.*

In higher dimensions, objects are points and regions are half-spaces. If there is no hypothesis on the data, $f_0(k)$ is bounded by the worst case size of a convex hull in dimension d, that is $O(n^{\lfloor \frac{d}{2} \rfloor})$

THEOREM 3.30 *The influence graph maintains the convex hull of n points in dimension d in expected time $O(\log n)$ per insertion if $d \leq 3$ and $O(n^{\lfloor \frac{d}{2} \rfloor - 1})$ per insertion if $d \geq 4$. The order of insertion is assumed to be random.*

Dynamic algorithm. The above algorithm is semi-dynamic, points can be added but not removed. The difficulty of dynamizing convex hull is that during the deletion of a point many other points may appear on the hull. Thus it is necessary to store all points, even those which are not on the current convex hull. In [Clarkson et al., 1993] a solution based on the history of insertion technique is given:

THEOREM 3.31 *The convex hull of n points in dimension d can be dynamically maintained with expected update time in $O(n^{\lfloor \frac{d}{2} \rfloor - 1})$ if $d \geq 4$ and in $O(\log n)$ if $d \leq 3$. The points are assumed to be inserted in a random order, and a deleted point can be any equally likely.*

9.3.2   VORONOI DIAGRAMS AND DELAUNAY TRIANGULATIONS

Points. We recall the definition of the Voronoi diagram in the plane. Given a set of points in the plane called sites, the Voronoi diagram is the partition of the plane such that a cell is composed of the set of points which are closer to one site than to the others. The dual graph of the Voronoi diagram, the Delaunay triangulation, is obtained by linking two sites if their Voronoi regions are adjacent.

A characteristic property of the Delaunay triangulation is that the circle circumscribing a triangle does not contain any site, or in the dual,

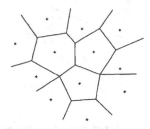

Figure 3.13.   Voronoi diagram

the center of the circle is at the same distance from three points (triangle vertices) and closer to these sites than to any other sites. Therefore objects are sites, regions are triangles and a point conflicts a triangle if the point is inside the circle circumscribing the triangle so that the Delaunay triangulation corresponds to the triangles without conflicts. Then the result follows [Boissonnat et al., 1992]. The algorithm has been fully dynamized in [Devillers et al., 1992] using the history of insertions.

THEOREM 3.32 *The influence graph maintains the Delaunay triangulation and the Voronoi diagram of n sites in the plane in $O(log n)$ expected insertion time, $O(log log n)$ expected deletion time if the insertion sequence is randomized and if the deleted site any equally chosen present site.*

THEOREM 3.33 *The influence graph maintains the Delaunay triangulation and the Voronoi diagram of n sites in dimension $d \geq 3$ in $O(n^{\lfloor \frac{d+1}{2} \rfloor} - 1)$ expected insertion time if the insertion sequence is randomized.*

Line segments.   Voronoi diagrams can also be defined for other kind of sites than points. The most frequent case is the Voronoi diagram of line segments or polygons (see Figure). The influence graph technique applies here also: the objects are line segments, regions are associated to the edges of the Voronoi diagram, which are defined by four objects, an object conflicts a region when the object intersects the union of circles centered on the Voronoi edge and touching the closest obstacle (see Figure).

9.3.3   ARRANGEMENTS

Another classical problem in computational geometry is the computation of the arrangement of curves and surfaces, that is the subdivison of the plane or the space induced by a set of curves or surfaces.

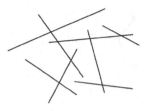

Figure 3.14.   Arrangement of line segments

Arrangements of line segments.   One of the simplest versions of
this problem is "given a set of n line segments in the plane, compute all
intersection points". Remark that the number of intersection points t
may vary between 0 and n^2.

The influence graph yields to a simple algorithm which is optimal on
average. Describing the algorithm reduces to describing the influence
graph. The algorithm constructs in fact the trapezoidal map of the
arrangement: for each segment end-point and each intersection point,
two vertical rays are drawn, and stopped at the first encountered line
segments above and below. If objects are line segments, regions are
trapezoids defined by at most four segments and a segment conflicts a
trapezoid if they intersect. Then the trapezoids of the map are exactly
the regions without conflicts. The influence graph technique applies.
Here $f_0(r)$ is r plus the average number of intersection points in a random
sample of size r. An intersection point of the whole arrangement appears
in the sample if the two segments defining it have been chosen: the
probability is $(\frac{r}{n})^2$. Therefore $f_0(r) = r + \frac{r^2}{n^2}t$ and the result is as follows:

THEOREM 3.34   *The influence graph maintains the trapezoidal map of
an arrangement of n segments in expected insertion time $O(logn + \frac{t}{n})$ if
the insertion sequence is randomized. t is the size of the arrangement.*

Arrangements of triangles.   The natural generalization of the preceding work to the three dimensional case consists in using vertical walls
passing through triangles edges and through intersection of triangles (see
[Boissonnat and Dobrind, 1992] for details and results).

THEOREM 3.35   *The influence graph computes the lower envelope of n
triangles in expected time $O(n^2\alpha(n)logn)$ if the insertion sequence is
randomized (α is the inverse of the Ackermann function [Abramowitz
and Stegun, 1965] and is extremely slowly growing).*

REFERENCES

This chapter is based on the articles [Chauvin et al., 1999, Cerf, 1994, Devillers, 1996, Jonassen and Knuth, 1978, Geniet et al., 1996, Laroche et al., 1999, Louchard and Schott, 1991, Louchard et al., 1997, Maier, 1991b, Maier and Schott, 1993a, Maier and Schott, 1993b, Régnier, 1989, Rhee and Talagrand, 1987, Rösler, 1989]

Chapter 4

SOME APPLICATIONS IN SPEECH RECOGNITION

1.   INTRODUCTION

The speech signal is the result of a voluntary and well-controlled movement of the speech apparatus. The design of an adequate modeling of speech patterns has been a constant concern since the beginning of automatic speech recognition research. In template methods [Sakoe and Chiba, 1978], the acoustic variability modeling of vocabulary consisted in storing several references for the same lexical unit, or in deriving typical sequences of acoustic frames by resorting to some kind of averaging method. These solutions were rather inefficient and expensive, even though they can provide a viable solution for a variety of applications. The idea of the statistical modeling of the spectral properties of speech gave a new dimension to the problem. The underlying assumption for all statistical methods is that speech can be adequately characterized as a random process whose parameters can be estimated effectively. We consider that the intelligent movement of the vocal track controlled by a speaker uttering a word in a specific acoustic and prosodic context is a temporal sequence of random events emitted by a Markov source. We therefore adopt a Bayesian attitude and quantify the uncertainty about the occurrence of an event by a probability.

The most widely used statistical method is the hidden Markov model (HMM) approach, first implemented for speech recognition during the seventies [Baker, 1975, Jelinek, 1976]. The basic HMM model has led to very good performances in various domains of speech recognition [Levinson, 1986, Schwartz and Austin, 1991, Normandin et al., 1994, Bahl et al., 1993].

153

In this chapter, we present a specific class of HMM, the second-order HMM (HMM_2), that generalize the first-order HMM. We also present some data analysis methods to classify and reject singular utterances that HMM_1 can not discriminate. The intrinsic limitations of this model will be progressively pointed out, as well as the necessity of incorporating into the model some knowledge about the speech communication process.

To summarize, this chapter shows how stochastic modeling and data analysis can be used in speech recognition systems, especially for a better representation of time varying phenomena.

2.   ACOUSTIC-FRONT END PROCESSOR

The articulatory characteristics of the human phonatory system indicate that the speech signal is quasi stationary for periods of 30-50 ms. The acoustic-front end processor acts as a data compressor whose purpose is to divide speech input into blocks (called *frames*) and to give a representation of these frames into the d-dimensional space $\mathcal{R}^d$.

In 1980, Davis and Mermelstein [Davis and Mermelstein, 1980] compared a comparison of several parametric representations of speech and concluded that *Mel-Frequency Cepstrum* Coefficients (MFCC) resulted in the best performance.

In 1982, Chollet [Chollet and Gagnoulet, 1982] gave the algorithm and the code in FORTRAN of a reference acoustic front end processor that has been widely adopted. This processor computes the cepstrum coefficients as follows:

- The acoustic signal is low-filtered, sampled at a frequency at least twice the cutoff filter's frequency, and blocked into overlapping windows.

- In each window, we compute a short-term spectrum, usually by applying a Fast Fourier Transform (FFT) algorithm. For MFCC computation, the power spectrum is filtered by a battery of triangular filters spaced on a Mel-frequency scale.

- The log energies of each filter constitute the output of a filter bank and are the input of an inverse discrete cosine transform that plays the role of the inverse FFT.

In a standard non-noisy speech recognition system, the typical values used are: 7 kHz for low-pass filtering, 16 kHz as the sampling rate, 25 ms as the window length and 8 ms as the frame shift. A frame is represented by a vector of 12 MFCC coefficients. The first coefficient represents the energy.

To model the spectrum shape, and to take into account the correlation between successive frames, we compute dynamic features based on the first and second-order derivatives of the MFCC vectors. Furui [Furui, 1986] proposes to compute the dynamic features using a regression polynomial:

$$a_m(t) = \frac{\sum_{n=-n_0}^{n_0} n x_m(t+n)}{\sum_{n=-n_0}^{n_0} n^2} \qquad (4.1)$$

where $x_m(t+n)$ is the m $(m = 0, \ldots 11)$ cepstrum coefficient of frame $t + n$. The $2n_0 + 1$ frames involved in the computation of $a_m(t)$ are centered around frame t. $a_m(t)$ is the mth first-order dynamic coefficients of frame t. Interested readers may refer to the work of Furui [Furui, 1981, Furui, 1986] and Hanson & Junqua [Hanson and Applebaum, 1990, Junqua et al., 1995b] for an extensive view of these methods. Finally, speech may be compressed into a real-valued 2-dimensional array in which a column represents a frame and a row represents a static or a dynamic coefficient.

In automatic speech recognition, we have to tackle several issues due to the variability of the input signal. As soon as the speech has been produced, distortion occurs. This can be the result of the channel of communication (e.g. a telephone line), or more simply, of an undefined microphone, both of which introduce a convolutionary noise.

This distortion acts like a filter whose influence can be removed by applying a technique called *cepstral mean subtraction* or *cepstral mean normalization* (CMN). Several authors [Junqua et al., 1995b, Junqua et al., 1995a] have shown that subtracting the mean frame, computed on a long term basis, improves the performances significantly.

3.   HMM SPEECH MODELING

The HMM can represent various speech units: from isolated or connected words such as sequences of digits to connected phonemes spontaneously uttered. Using the notations and definitions introduced on page 16, we can specialize the HMM by giving the definition of a state.

In a classical speech recognition system, the HMMs model the variability of the speech signal by associating each state s_i with a mixture of Gaussian distributions:

$$b_i(O_t) \overset{\text{def}}{=} \sum_{m=1}^{M} c_{im} \mathcal{N}(O_t; \mu_{im}, \Sigma_{im}) \quad \text{with} \quad \sum_{m=1}^{M} c_{im} = 1 \qquad (4.2)$$

where O_t is the input vector (the frame) at time t and $\mathcal{N}(O_t; \mu, \Sigma)$ the expression of the likelihood of O_t using a multi-dimensional Gaussian

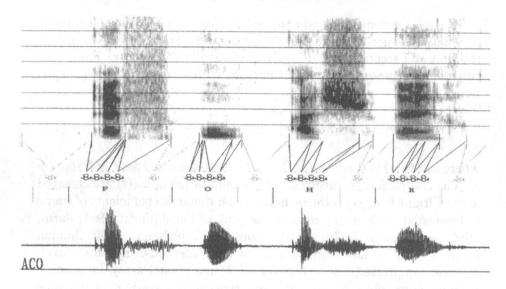

Figure 4.1.   Each state captures a speech segment.

density with mean μ and covariance Σ. See the C++ programs that perform this computation (appendix page 205).

HMM-based speech modeling assumes that the input signal can be split into segments modeled as states of an underlying Markov chain and that the waveform of each segment is a stationary random process as shown in Figure 4.1. In a first-order hidden Markov model, the sequence of states is assumed to be a first-order Markov chain. This assumption is mainly motivated by the existence of efficient and tractable algorithms – EM and Viterbi algorithms – for model estimation and recognition, whose complexity is proportional to n^2t where n is the number of states of the model and t the utterance length. The EM algorithm acts as a training algorithm, whereas the Viterbi algorithm extracts the best alignment between the states and the frames, computes its probability and determines the recognized unit.

However, HMM_1 suffer from several drawbacks. For instance, in an HMM_1 the frames inside a segment are assumed to be independent, and trajectory modeling (*i.e.* frame correlation) in the frame space is not included. By incorporating short-term dynamic features to model spectrum shape, HMM_1 can be made to overcome this drawback. Modeling segment duration, which follows a geometric law as a function of time, remains another major drawback of HMM_1. In this chapter, we investigate models where the underlying state sequence is a second-order Markov chain, and we study their capabilities in terms of duration and frame correlation modeling. The major disadvantage of our approach being the computational complexity, we propose an appropriate implementation of the re-estimation formulae that yields algorithms only n_i times slower compared to HMM_1, where n_i is the average input branching factor of the model. We show that HMM_2 overcomes HMM_1, but the advantage is significantly reduced after a post-processing in which durational constraints based on states occupancy is incorporated into an HMM_1-based system.

4. SECOND-ORDER HMMS

Second-order HMMs have been already defined in a previous chapter (see page 17). Their major advantages are their capabilities in duration modeling as explained in Equation 1.5.

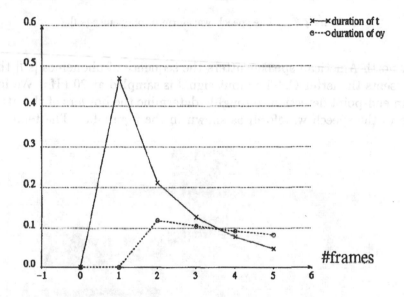

Figure 4.2. Distributions of duration on the central state of the phonemes /t/ and /oy/. The duration is expressed in term of frames whose duration is 8 ms. The central state of the HMM_2 modeling the phoneme /oy/ has a transition probability $a_{j-1jj+1}$ equal to 0.

As shown in Figure 4.2, such a model provides a better fit for the true segment duration. In this figure, we have represented the duration distributions in the middle state of a 3-state phone-based HMM_2 for two models: phoneme 't' and 'oy' fluently spoken by a native American. 't' is a short phoneme and its distribution, as modeled by an HMM_2, is left-concentrated, whereas the distribution of the long diphthong 'oy' is flatter and shifted to the right. Equations 1.5 govern the duration model. The central state of the HMM_2 modeling the phoneme /oy/ has a transition probability $a_{j-1jj+1}$ equal to 0. Therefore, a second-order HMM can eliminate singular alignments given by the Viterbi algorithm in the recognition process, when a state captures just one frame and all other speech frames fall into the neighboring states.

4.1   EXAMPLE OF A SINGULAR ALIGNMENT IN A HMM_1

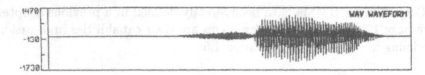

Figure 4.3.   The speech waveform of phonemes [s ɪ].

A north American speaker utters the sequence of phonemes [s ɪ] that represents the letter C. The input signal is sampled at 20 kHz. We first do an end-point detection to roughly determine the borders of the utterance in the speech waveform as shown in the figure 4.3. The temporal

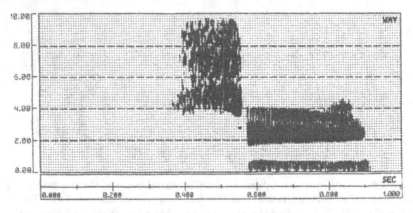

Figure 4.4.   Typical spectrogram of phonemes [s ɪ]. The vertical axis of the spectrogram ranges from 0 and 10 kHz.

signal and the spectrogram (Fig. 4.4) clearly show the two phonemes: /s/ et /ɪ/.

The Viterbi algorithm is used on two HMM_1: one associated with the letter C (the right one) and one associated with the letter T (the wrong one which will be recognized). For each model λ, we plot the curve PROB_λ that represents $\log(\delta_t(j))$ as a function of t along the best path. The vertical axis ranges from 30 and -250.

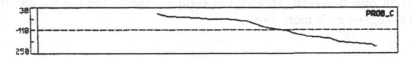

Figure 4.5. Variations of $\log(\delta_t(j))$ along the best path in the Viterbi algorithm.

For model C (Fig. 4.5), the value $\log(\delta_t(j))$ – the probability of the partial alignment (cf. Equation. 1.1, page 15) – decreases slowly. The final value is roughly -230.

We next represent the sequence of states belonging to the best path: PATH_λ.

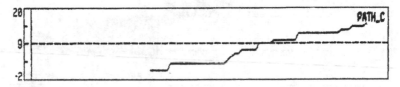

Figure 4.6. $\log(\delta_t(j))$ assuming model C.

The figure 4.6 represents the state sequence of the whole alignment. In this figure, the vertical axis represents the 15 states of a model. The two initial states of the model have captured the fricative part of the utterance represented by phoneme /s/.

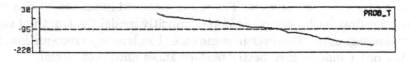

Figure 4.7. $\log(\delta_t(j))$ assuming the model T.

We now use the Viterbi algorithm for model T (Fig. 4.7). The final value of $\log(\delta_t(j,k))$ reaches roughly -210 which is a better score as shown in Fig PROB_T. Therefore, model T is recognized.

Let's look at Figure 4.8 which shows the best path that the Viterbi algorithm has found using model T. The process stays for an outrageously long time on the first state – roughly one third of the whole duration – and we see that the first state of model T is mapped on the phoneme /s/. The next states share the remaining time. During the training of HMM T, the first state has been automatically specialized to capture the burst of phoneme /t/. This acoustic cue has the same spectral shape as phoneme /s/ but a shorter duration. Because HMM_1 are insensitive to segment duration – the shorter the segment, the higher the probability – this alignemnt is more likely.

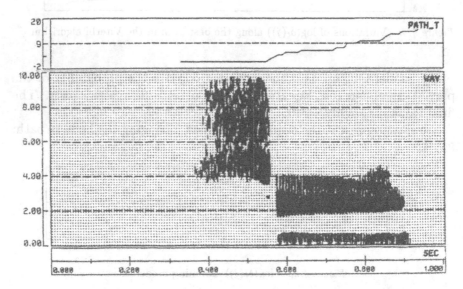

Figure 4.8. The best path given by the Viterbi algorithm assuming HMM T.

It is easy to modify the Viterbi algorithm to incorporate some constraints on state duration, but this algorithm does not remain optimal from a Markov point of view. The use of a second-order model can alleviate this drawback because HMM_2 explicitly model the length of short duration segments. However in some cases like language recognition, the higher order models give better performances [du Preez, 1998].

5.   THE EXTENDED VITERBI ALGORITHM

The extension of the Viterbi algorithm to HMM_2 is straightforward. We simply replace the reference to a state in the state space **S** by a reference to an element of the 2-fold product space **S x S**. The most likely state sequence is found by using the probability of the partial alignment ending at transition (s_j, s_k) at times $(t-1, t)$:

$$\delta_t(j,k) \stackrel{\text{def}}{=} P(q_1, ...q_{t-2}, q_{t-1} = s_j, q_t = s_k, O_1, ..., O_t | \lambda), \quad (4.3)$$
$$2 \leq t \leq T, \quad 1 \leq j, k \leq N.$$

Recursive computation is given by:

$$\delta_t(j,k) = \max_{1 \le i \le N}[\delta_{t-1}(i,j) \cdot a_{ijk}] \cdot b_k(O_t), \qquad (4.4)$$

$$3 \le t \le T, \quad 1 \le j,k \le N.$$

In continuous speech recognition, the borders between the phonemes are unknown. We connect the HMM in a wider network and define a wider HMM which is the linguistic network in which the Viterbi algorithm can be run. The segmentation and the labelling of the utterance are determined from the best-state sequence. A real case is shown in Figure 4.9. In this example, we have modeled the 35 French phonemes by 35 HMM_2. The spoken sentence is: *C'est en forgeant qu'on devient forgeron.* The phonemes's labels are given vertically. Refer to Table 4.1(from Calliope [Calliope, 1989]) from the meaning of each phoneme.

Consonants

[p]	paie	cl p	[t]	taie	cl t	[k]	quai	cl k
[b]	baie	vcl b	[d]	dais	vcl d	[g]	gai	vcl g
[m]	mais	m	[n]	nez	n	[ɲ]	gagner	
[f]	fait	f	[s]	sait	s	[ʃ]	chez	S
[v]	vais	v	[z]	zéro	z	[ʒ]	geai	Z
[ω]	ouais	w	[j]	yéyé	j	[l]	lait	l
[R]	raie	R						

Vowels

[i]	lit	i	[y]	lu	y	[u]	loup	u
[e]	les	e	[ø]	leu	2	[o]	lot	o
[ɛ]	lait	E	[ɔ]	lotte	O	[a]	là	a
[ə]	le	@	[ɛ̃]	lin	in	[ã]	lent	an
[õ]	long	on						

Table 4.1.   The French phonemes.

The pixels represent the likelihood of the frames computed using the *pdf* that define the states of the linguistic network.

6. THE EXTENDED BAUM-WELCH ALGORITHM

To adjust the second-order HMM parameters, we define the new forward and backward functions on which the extended Baum-Welch algorithm is based. The forward function $\alpha_t(j,k)$ defines the probability of the partial observation sequence, $O_1, ..., O_t$, and the transition (s_j, s_k) between time $t-1$ and t:

$$\alpha_t(j,k) \stackrel{\text{def}}{=} P(O_1, O_2, ..., O_t, q_{t-1} = s_j, q_t = s_k|\lambda), \quad 2 \le t \le T, \quad (4.5)$$

$$1 \le j,k \le N.$$

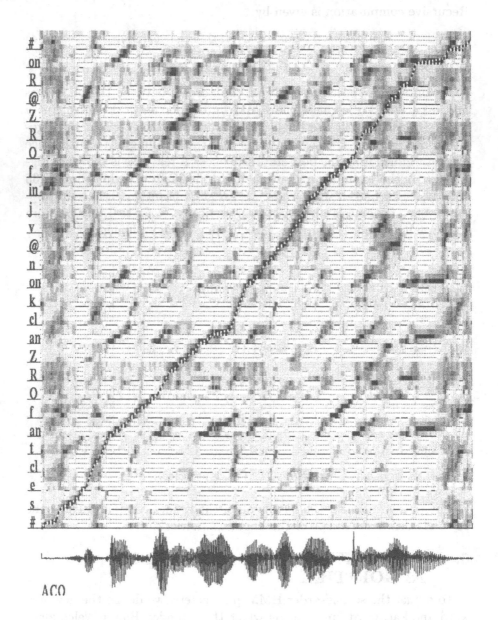

Figure 4.9. [
Example of a time warping alignment found by the Viterbi algorithm]The best path found by the Viterbi algorithm determines a segmentation in phonemes of the sentence : *C'est en forgeant qu'on devient forgeron.* Refer to Table 4.1 for the meaning of the Y-axis.

In an HMM_2, $\alpha_t(j,k)$ can be computed from $\alpha_{t-1}(i,j)$ in which (s_i, s_j) and (s_j, s_k) are two transitions between states s_i, s_j and s_k:

$$\alpha_{t+1}(j,k) = \sum_{i=1}^{N} \alpha_t(i,j).a_{ijk}.b_k(O_{t+1}), \quad 2 \le t \le T-1, \quad 1 \le j,k \le N.$$

$$(4.6)$$

Assuming the same restrictions on the final state of the Markov chain (cf. page 14), the probability of the whole sequence of observations is defined as follows:

$$P(O_1, O_2, ..., O_t|\lambda) = \sum_{i=1}^{N} \alpha_T(i,N). \qquad (4.7)$$

Similarly, the backward function $\beta_t(i,j)$, defined as the probability of the partial observation sequence from t+1 to T, given the model λ and the transition (s_i, s_j) between times t-1 and t, can be expressed as :

$$\beta_t(i,j) \stackrel{\text{def}}{=} P(O_{t+1}, ...O_T|q_{t-1} = s_i, q_t = s_j, \lambda), \quad 2 \le t \le T-1, (4.8)$$
$$1 \le i,j \le N.$$

The backward function is computed recursively as follows:

$$\beta_t(i,j) = \sum_{i=1}^{N} \beta_{t+1}(j,k).a_{ijk}.b_k(O_{t+1}), \quad 2 \le t \le T-1, \quad 1 \le j,k \le N.$$

$$(4.9)$$

Given a model λ and an observation sequence O, we define $\eta_t(i,j,k)$ as the probability of the transition $s_i \longrightarrow s_j \longrightarrow s_k$ between $t-1$ and $t+1$ during the emission of the observation sequence.

$$\eta_t(i,j,k) \stackrel{\text{def}}{=} P(q_{t-1} = s_i, q_t = s_j, q_{t+1} = s_k/O, \lambda), \quad 2 \le t \le T-1.$$

We deduce:

$$\eta_t(i,j,k) = \frac{\alpha_t(i,j)a_{ijk}b_k(O_{t+1})\beta_{t+1}(j,k)}{P(O|\lambda)}, \quad 2 \le t \le T-1. \qquad (4.10)$$

As in the first order, we define $\xi_t(i,j)$ and $\gamma_t(i)$:

$$\xi_t(i,j) \stackrel{\text{def}}{=} \sum_{k=1}^{N} \eta_t(i,j,k), \qquad (4.11)$$

$$\gamma_t(i) \overset{\text{def}}{=} \sum_{j=1}^{N} \xi_t(i,j). \tag{4.12}$$

$\xi_t(i,j)$ represents the *a posteriori* probability that the stochastic process accomplishes the transition $s_i \to s_j$ between $t-1$ and t assuming the whole utterance. In Figure 4.10 we have represented these probabilities by grey levels. The observations come from the training sentence *"et le type de musique jouée"*. This figure is similar to Figure 4.9 which shows the best path resulting from the Viterbi algorithm. In many training situations, the Viterbi algorithm is used in the place of the Baum-Welch algorithm because it is simpler to implement. The value of 1 is set to $\eta_t(i,j,k)$ if i,j,k are 3 states on the best path at times $t-1, t, t+1$. Also, this example illustrates the approximation performed by the Viterbi algorithm which computes the probability of a sequence of observations using only the best alignment rather that summing all alignments.

$\gamma_t(i)$ represents the *a posteriori* probability that the process is in the state i at time t assuming the whole utterance.

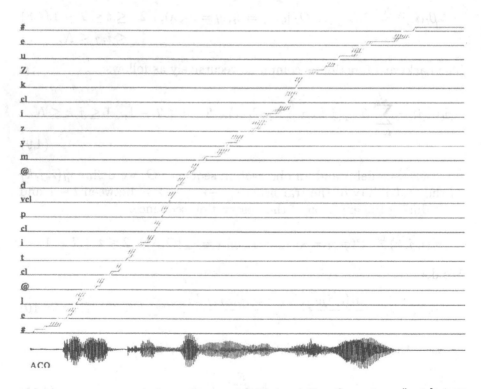

Figure 4.10.   a posteriori transition probability assuming the sentence:" et le type de musique jouée".

At this point, to get the new maximum likelihood estimation (ML) of the HMM_2, we can choose two ways of normalizing: one way gives an HMM_1, the other an HMM_2.

The transformation in HMM_1 is done by averaging the counts $\eta_t(i,j,k)$ over all the states i that have been visited at time $t-1$.

$$\eta_t^1(j,k) \overset{\text{def}}{=} \sum_{i=1}^{N} \eta_t(i,j,k) \qquad (4.13)$$

is the classical first order count of transitions between 2 HMM_1 states between t and $t+1$.

Finally, the first-order maximum likelihood (ML) estimate of $\overline{a_{ijk}}$ is:

$$\overline{a_{ijk}} = \frac{\sum_t \eta_t^1(j,k)}{\sum_{k,t} \eta_t^1(j,k)} = \frac{\sum_{i,t} \eta_t(i,j,k)}{\sum_{i,k,t} \eta_t(i,j,k)}. \qquad (4.14)$$

This value is independent of i and can be written as $\overline{a_{jk}}$.

The second-order ML estimate of $\overline{a_{ijk}}$ is given by the equation:

$$\begin{aligned}
\overline{a_{ijk}} &= \frac{\sum_t \eta_t(i,j,k)}{\sum_{k,t} \eta_t(i,j,k)} \\
&= \frac{\sum_{t=1}^{T-2} \eta_{t+1}(i,j,k)}{\sum_{t=1}^{T-2} \xi_t(i,j)}.
\end{aligned} \qquad (4.15)$$

In a continuous HMM_2, as defined page 155, the observations are drawn from mixtures of normal laws. The ML estimates of the mean and covariance are given by the formulas:

$$\overline{\mu_i} = \frac{\sum_t \gamma_t(i)O_t}{\sum_t \gamma_t(i)}, \qquad (4.16)$$

$$\overline{\Sigma_i} = \frac{\sum_t \gamma_t(i)(O_t - \mu_i)(O_t - \mu_i)^t}{\sum_t \gamma_t(i)}. \qquad (4.17)$$

The C++ function addWeightedSample() (cf. appendix page 205) performs this computation on the Gaussian $\mathcal{N}(\mu_i, \Sigma_i)$.

7.   IMPLEMENTATION AND COMPLEXITY

A naive implementation of the recursion for the computation of α and β requires on the order of N^3T operations, compared to N^2T for the standard HMM_1. A more precise analysis of Equation(4.6) shows that the sum is taken over all non-zero transitions a_{ijk} for a given pair of successive states (j,k). In a framework of an object oriented language like C++, a transition is represented by a class:

```
struct Transition_t
{
  short int rgij ;        // index of pair (origin, middle)
  short int rgjk ;        // index of pair (middle, extremity)
  short int origin, middle, extremity ; // i -> j -> k
  float  prob ;           // transition probability
...
} ;
```

All the transitions of the HMM_2 are stored in a list:

```
Transition_t *Transitions;
```

that will be dynamically allocated as soon as the number of transitions nTransitions is known. The iterations on this list are performed using a special object called *an iterator*. Direct access to a specific pair of states is not necessary. The variable trit belongs to the class Transiter and implements an iterator in the transition list of HMM oldHmm.

```
TransIter trit(oldHmm) ;
Transition_t trans ;
while (trit.next(trans))
{
  ... // process item trans
}
```

For example, Equation 4.6 can be computed in an efficient way:

```
TransIter trit(oldHmm);
Transition_t trans1;
while (trit.next(trans1))
{
  alpha[it][trans1.rgjk] += alpha[it-1][trans1.rgij]
        * trans1.prob * oldHmm.b1mu(trans1.extremity,frm);
  scaleFactor = MAX(scaleFactor,alpha[it][trans1.rgjk]);
}
```

The call b1mu(trans1.extremity, frm) represents a call to the method that performs the computation of the likelihood of frame frm at state trans1.extremity in the Hmm oldHmm. Remember that we want to compute a new estimate of Hmm from an old one. The meanings of the variables scalefactor and scale[it] will be explained below.

The sentence probability probS is computed using the values of α at time t – the last value of it – on each pair of states that reaches the final state – called lastNode of the Hmm. The array of flags finalPair indicates if the corresponding pair is final or not. nPairs represents the number of pairs of successive states in the HMM. The average input branching factor N_{input} of the model may be defined as:

$$N_{input} \stackrel{def}{=} \frac{nPairs}{N}$$

```
    double probS = 0.;
// where are the distinct pair ending in lastNode?
    for (int i = 0 ; i < nPairs ; i++)
        if (finalPair[i]) probS += alpha[t][i];
```

This implementation suggests that there is roughly a factor of N_i in terms of space and computation requirements between HMM_1 and HMM_2. In our system, the average input branching factor is 2 and the re-estimation and recognition algorithm are less than two times slower, which is still tractable. In fact, experimental measures on CPU times of both systems have shown that HMM_2 are only 1.5 slower than HMM_1, since the system spends half the time in computing the likelihood of frames given a state output *pdf* with full covariance matrices than accessing the lists of transitions.

7.1   BINARY REPRESENTATION ISSUES

A straightforward implementation of Equation 4.6 for the computation of α is intractable. Underflows occur rapidly. Therefore, we have to introduce a scale factor in the definition of α and β to keep their values in the range of allowable representations (roughly $[10^{-34}, 10^{34}]$).

During the computation of α, at each time it, the highest probability is raised to 1, the scale factor is first stored in the variable scalefactor and next in the array scale[ ].

```
    for (int i = 0 ; i < nPairs ; i++)
        alpha[it][i] /= scaleFactor; // normalize

    // save ct
    scale[it] = scaleFactor;
```

Hopefully, the scale factors can be removed in Equation 4.10 because the contributions of the scale factors stored in scale[] are the same in the numerator and denominator of Equation 4.10.

8.   DURATION MODEL

Even if the duration of a segment is better modeled by two parameters in an HMM_2, thus avoiding singular state assignment as previously mentioned, it is nevertheless necessary to implement duration constraints based on the relative duration of the segments corresponding to successive states as in [Suaudeau and André-Obrecht, 1993]. The reason is that most of the errors of our HMM_2-based word recognition system come from singular alignments given by the Viterbi algorithm. We observed that state durations were strongly correlated for states in a model. In order to take this correlation into account, we have specified a set of classes of correct alignments on a one class per model basis. Given an utterance, an alignment between a model and the utterance is defined by a vector of relative duration of the states of the model. This alignment is found using the Viterbi algorithm. We denote by:

- **w**, a d-frame long word which has been aligned with HMM λ. Each state i among the N states of λ captures d_i frames. If all states must be visited, we have: $d = d_1 + d_2 + ... + d_N$

-
$$x \overset{\text{def}}{=} (d, \frac{d}{d_2}, \frac{d}{d_3}, ..., \frac{d}{d_N}). \tag{4.18}$$

- g_λ, the mean vector associated with the class of λ, and V_λ the covariance matrix

- $\det V_\lambda^{1/N}$, a normalizing factor that ensures that all matrices have a determinant equal to one.

Given a word model and the class of correct alignments of this model, we measure the distance between alignments using the Mahalanobis distance:

$$d^2(x, g_\lambda) \overset{\text{def}}{=} \det V_\lambda^{1/N}(\mathbf{x} - \mathbf{g}_\lambda)^t \mathbf{V}_\lambda^{-1}(\mathbf{x} - \mathbf{g}_\lambda). \tag{4.19}$$

This distance weights the probability of the Viterbi's alignment during a post-processing step where the N-best answers given by the recognition algorithm [Schwartz and Austin, 1991] are re-scored. The N-best recognition algorithm generalizes the Viterbi algorithm and extracts the N best alignments.

$$\text{FinalScore} = A \cdot d^2(x, g_\lambda) + B \cdot log(P(O|\lambda)). \tag{4.20}$$

A and B are normalizing constants determined empirically on the training set.

9. EXPERIMENTS ON CONNECTED DIGITS
9.1 TEST PROTOCOL

First-order HMM and second-order HMM have been comparatively assessed using the same database of digits, *i.e.* the adult part of the TI-NIST database.

The TI-NIST database has eleven digits: "zero", "one", ... , "nine", "oh" recorded by 225 adults (111 males and 114 females). Each speaker has spoken 77 strings, of which 22 are isolated digits, and the remaining strings fall into 5 groups of 11 strings of length 2, 3, 4, 5 and 7. The vocabulary is made up of 23 models, one per digit and gender, and one for the background noise. The state output densities are mixtures of 9 Gaussian estimates with full covariance matrices. For the comparison, we have used models with the same topology and same number of *pdf*. In particular, digit models have 6 states with 5 self loops and no skip transition, whereas the background noise model has only 2 states and one self loop.

9.2 ACOUSTIC ANALYSIS

Using a frame shift of 12 ms and a 25 ms window, 12 cepstral coefficients corresponding to an approximate Mel-frequency warped spectrum are computed. The first coefficient, called loudness, was removed. Two analysis feature vectors incorporating dynamic features have been specified in order to explore the capability of HMM_2 to capture frame correlations:

- 24 coefficients: 11 static, 12 dynamic first-order coefficients plus the second-order energy coefficient $\Delta\Delta E$.

- 35 coefficients: 11 static, 12 dynamic first-order coefficients plus 12 dynamic second-order coefficients.

9.3 COMPARISON HMM_1 / HMM_2

Tables 4.2 and 4.3 summarize the recognition results. In these tables, the string error rates and the corresponding 95% confidence intervals are given. Table 4.4 gives the results at the word level. In the different experiments, we used the 8700 strings from the test part of the TI-NIST database containing 28383 digits. Throughout these experiments, a grammar of unknown string length was used to specify the linguistic network.

Three major conclusions can be drawn from these results:

Parameter	Male + Female	
	HMM_1	HMM_2
24	4.5% (4.1 5.0)	2.4% (2.1 2.7)
35	3.7% (3.3 4.1)	2.4% (2.1 2.7)

Parameter	Male + Female	
	HMM_1	HMM_2
24	2.8% (2.5 3.2)	2.2% (1.9 2.5)
35	2.3% (2.0 2.6)	2.1% (1.8 2.4)

Table 4.2. String error rates (without post-processing)

Table 4.3. String error rates (with post-processing)

	HMM_1	HMM_2
Insertions	174	159
Deletions	14	20
Substitutions	31	34
String error rate	2.3 %	2.4 %
% correct	99.8	99.8
Accuracy	99.2	99.2

Table 4.4. Comparison between HMM_1 (with post-processing) and HMM_2 (without post-processing)

1. HMM_2 outperforms HMM_1 in the absence of post-processing, and HMM_2 without post-processing is almost equivalent in performance to HMM_1 with post-processing.

2. Acceleration coefficients do not significantly improve performance, especially with HMM_2.

3. The offset in performances is greatly reduced when a post-processor is used to take into account the duration constraints.

Point 1 can be explained by the capability of HMM_2 to explicitly model the event that a state can be visited just one time, and to eliminate singular alignments given by the Viterbi algorithm in the recognition process when a state captures just one frame, whereas all other speech frames fall into the neighboring states. Thus, the trajectory of speech, in terms of state sequence, is better modeled by HMM_2.

Point 2 has already been mentioned in relation to clean speech and HMM_1 models [Hanson and Applebaum, 1990].

Since the beginning of this study in 1990, several systems have produced better performances on the TI-NIST corpus [Haeb-Umbach et al., 1993, Cardin et al., 1993]. These systems involve sophisticated acoustic analysis and training techniques. Our word recognition system, based on HMM_1 models, which serves as the reference system to which the HMM_2-based system was compared, gives results similar to the system described by Wilpon in 1993 [Wilpon et al., 1993], *i.e.* 2.4 % string error rate with a 10 state model with 9 Gaussian *pdf* per mixture and telephone bandwidth speech. In our system, we have 6 states per model but 2 models per digit. This keeps the number of parameters fairly constant.

Usually, the duration constraints are implemented in post-processors that are trained separately. The maximum likelihood principle in the training is thus not guaranteed, whereas HMM_2s converge with the forward-backward algorithm naturally to Ferguson-like models in which the state duration is represented by a non parametric *pdf* for small values and a geometric law for higher values. Besides, the duration modeling with parametric *pdf* in a semi-Markov process [Levinson, 1986] results in a significant increase in computational complexity, whereas a second-order Markov process gives a crude, but tractable, answer.

10.   EXPERIMENTS ON SPELLED NAMES OVER THE TELEPHONE

Automatic speech recognition of spelled names is a difficult task because of confusions between letters of the alphabet, distortions introduced by the telephone channel and variability due to an undefined telephone handset. The database used in our experiments is a subset of the speech telephone corpus collected at Oregon Graduate Institute (OGI) [Cole et al., 1992]. Over 4000 people called in response to public requests. They were prompted by a recorded voice to spell their first and last names, with and without pauses. Several speakers did not follow the instructions, and therefore we discarded their calls in the training and testing sets. For the training, we selected 60 alphabets and 1200 calls. For the testing, we selected 491 calls. As every speaker belongs to one set, the experiments conducted are speaker-independent.

10.1   ACOUSTIC FRONT-END

For the speech representation, we computed 12 static MFCC coefficients on a 32 ms window-length every 10 ms. We also concatenated 12 first-order regression coefficients to the static ones. The first coefficient, C_0, called loudness, was removed but we used the C_0 second-order regression. Additional second-order regression coefficients were not considered because we found in recent work [Junqua et al., 1995b] that they yield only a slight improvement.

10.2   PERFORMANCES

The comparison between HMM_1 and HMM_2 is summarized in Table 4.5. The low insertion rate in the HMM_2-based recognition system may be interpreted as a result of the additional constraints in state duration modeling that HMM_2s impose.

	HMM_1	HMM_2
Insertions	3.7	1.1
Deletions	1.4	2.4
Substitutions	11.8	11.4
% correct	86.8	86.1
Accuracy	83.1	85

Table 4.5.   Recognition accuracy (in %) for HMM_1 and HMM_2.

10.3   STATE TYING

In a continuously-spelled-name task, most of the confusions are due to the acoustic proximity of letters belonging to confusable subsets such as the E-set = {B, C, D, E, G, P, T, V }, A-SET = { A, H, J, K} or { L, M, N, S, X } depending on the vowel that is used to pronounce an isolated letter: /ɪ/ in the ESET, the diphthong /e i/ in the ASET or the vowel /e/ in {L, M, N, S, X}.

With one *pdf* per state, this proximity between models can be quantified state by state (cf. page 32) using the Kullback-Leibler (KL) measure between the *pdf* associated with each state of the HMM. Two corresponding states in two different HMMs belonging to a confusable subset, are tied together – they share the same *pdf* – if the measure is low enough and if the states capture the same vocalic sound.

The states were next split into several *pdf* by means of a top down Linde-Buzo-Gray algorithm (see the chapter devoted to statistical tools page 51, and the appendix page 214 in which this algorithm is given) followed by a bottom-up hierarchical agglomerating algorithm (cf. *the Ward algorithm* given page 47) based on a weighted-by-counts minimum increase of divergence criterion (see page 32).

10.4   THE SPECIFICATION OF HMMS WITH TIED STATES

In this section, we describe the supervised process of designing a set of HMMs having tied states in the framework of letter recognition. As we will see later, a human expert in phonetics plays a central role in this process. In some sense, the process of determining which state is candidate for tying is a way of incorporating some phonetic knowledge into the recognition process that implemented only probabilistic models until now.

All the HMMs have 6 states, except W which has 12. We call HMM^{α}, the HMM associated with letter α. s_i^{α} is the state i of the model HMM^{α}. The letters α and β are in the same class (E-SET, A-SET ...) if they are articulated with the same vowel (or diphthong). The process of specifying 26 HMMs with phonetic state tying is summarized in Algorithm 4.1.

The threshold R_0 controls the size of the lists. We determine it by studying the variations of the distance $d(s_i^{\alpha}, s_i^{\beta})$ computed on the 2 last states of a model drawn from the ESET and the model of /I/. The lower R_0, the less the states are tied.

In this algorithm, the state tying improves the discrimination between the models by incorporating some crude phonetic knowledge. Because the likelihood of the frame, given the tied state, is constant on all states sharing the same *pdf*, the Viterbi algorithm will not use this information to discriminate the models. Note that this technic should have avoid the singular alignment given page 159.

10.5   RESULTS

We trained 27 models (one for each letter, plus one for pauses) with an average of five *pdf* per state. All models had six states except W, which had 12 states. Recognition was carried out by means of a Viterbi algorithm constrained by a letter-pair grammar. Table 4.5 shows the results obtained without state tying. In this table, the accuracy is defined as: 100 - % insertions - % deletions - % substitutions. When state tying is used in the HMM specification, the performances raise to 89%. When gender-dependent HMM_2s were used, the letter recognition accuracy improved to 90%. State tying provides an improvement of roughly 4% in recognition accuracy. The main effect of state tying is to reduce the number of deletions.

Algorithm 4.1 The process of phonetic state tying on OGI.

Ligature (corpus, HMM^A, HMM^B, ... , HMM^Z)

input: a corpus of spelled letters.

output: the 26 HMM with tied states.

1. Learn one model per letter. Each state is represented by a normal *pdf*.

2. **for** i = 1 to 6 **do** /* there are 6 states in the HMMs */
   $L_i \leftarrow ()$ /* empty list */
     for $\alpha \in \{A, B, C, \dots, Z\}$ **do**
       for $\beta \in \{A, B, C, \dots, \alpha\}$ **do**
         compute the Kullback-Leibler's distance $d(s_i^\alpha, s_i^\beta)$
         if $d(s_i^\alpha, s_i^\beta) \leq R_0$ **and** α and β are in the same class **then**
           add s_i^α, s_i^β in the list L_i
         end
       done
     done
   done

3. Give these 6 lists to an expert who decides, for each pair, (s_i^α, s_i^β) if the state i of HMM^α and HMM^β can capture the same vocalic segment. This decision can be made by visualizing the Viterbi alignments (cf. Figure 4.6) or by examining the rank of the state in the model:start or end. The result given by the expert is a list from which the singular pairs of states have been removed. The low temporal and spectral resolutions of an HMM having one *pdf* per state are responsible for the occurrence of these singular pairs of states. This list determines which HMMs share the state i.

4. Run the Baum-Welch's algorithm to train the HMM.

5. Perform a Viterbi algorithm on the corpus to determine the pairs (frame, state) and run a clustering algorithm on each state, each cluster will be assumed Gaussian.

6. On each HMM, change the definition of the states; instead of having one *pdf* per state, use rather a mixture of *pdf* depending on the clusters found in step 5;

7. Perform a Baum-Welsh estimation to estimate the model parameters.

---

11.   EXPERIMENTS ON CONTINUOUS SPEECH

To further assess the modeling capabilities of HMM_2s, we developed a phone recognizer using the TIMIT [Garofolo et al., 1993] database and context-independent (CI) sub-word units. TIMIT is an acronym for *Texas Instruments* and *Massachusetts Institute of Technology*. The former has recorded the corpus, while the later has labeled the corpus in phonemes and words. The TIMIT corpus is a corpus of read speech designed to provide speech data for the acquisition of acoustic-phonetic knowledge and for the development and evaluation of automatic speech recognition systems. TIMIT contains a total of 6300 sentences, it also contains a training/testing subdivision. The test sub-corpus contains 50754 phonemes drawn from 1344 sentences pronounced by 112 men and 56 women. The confidence interval at a 5% risk is 0.4% for the actual recognition rates (around 70 – 75 %).

During the recognition experiments, a phone-based bigram was used. To compute the results, the set of 39 phonemes as defined in [Lee and Hon, 1989] was used. For the experiments, we used the training/test subdivision as specified by the TIMIT-CDROM. Training set: 8 sentences spoken by 462 speakers. Test set: 8 sentences spoken by 168 speakers,

11.1   ACOUSTIC FRONT-END

We computed 12 MFCC coefficients on a 32 ms window-length every 8 ms and we used the same set of acoustic features as in the spelled name recognition task over the telephone.

11.2   HMM SPECIFICATION

48 context-independent (CI) phonemes were trained with three states per model and an average of six *pdf*s per state. Left-to-right, self-loop and no-skip transition models were trained. Full covariance matrices were estimated at each state. For these experiments, tying was not used.

11.3   RESULTS

The recognition was carried out by means of a Viterbi algorithm constrained by a phone bigram grammar, computed on the training data. Results are given at the phone level. Table 4.6, where pauses were included in the scoring algorithm, summarize the results.

These results compare favorably with the 68% obtained in [Rathinavelu and Deng, 1995] with context-independent phone models. They

‖ $\Delta\Delta$	HMM_1	HMM_2 ‖
‖ Diagonal	65.4	65.7 ‖
‖ Full	70.5	70.6 ‖

(a) 35 coef. and 1111 *pdf*

‖ Δ	HMM_1	HMM_2 ‖
‖ Diagonal	65.4	65.6 ‖
‖ Full	69.4	69.4 ‖

(b) 24 coef. and 1329 *pdf*

Table 4.6.   Phone recognition rate on *TIMIT* using *HMM*

are also very close to those published in [Lamel and Gauvain, 1993]. In this paper, the authors obtained an accuracy of 69.1% with delta coefficients and 500 context-dependent HMMs. The use of context-dependent models with HMM_2 would probably contribute to enhanced performance. Also, these results show the influence of the correlation in the coefficients of the input frames. When full covariance matrices are used, we observe a 4 % increase in the results.

Chapter 5

SOME APPLICATIONS IN ROBOTICS

1. APPLICATION OF HIDDEN MARKOV MODELS

In this section, we describe a method based on hidden Markov models for learning and recognizing places in an indoor environment by a mobile robot. Hidden Markov models have been used for a long time in pattern recognition, especially in speech recognition. Their main advantages over other methods (*e.g.* neural networks, ...) are their capabilities to modelize noisy temporal signals of variable length.

We show in this section that this approach is well suited for learning and recognizing places by a mobile robot. Results of experiments on a real robot with distinctive places are given.

1.1 INTRODUCTION

The automatic recognition of places is an important issue that determines the capability of a mobile robot to localize itself in its environment. Place recognition is useful in a variety of tasks such as the automatic construction of topological maps [Kortenkamp and Weymouth, 1994]. Place learning and place recognition have been addressed by different approaches such as analytical methods or pattern classification methods. In the first approach, the problem is studied as a reasoning process. A knowledge-based system uses rules to build a representation of places. For instance, [Kortenkamp and Weymouth, 1994] uses rules about the variation of the sonar sensors to learn different types of places and adds visual information to recognize two places of the same type. [Yamauchi, 1995] uses ultrasonic sensors to build evidence grids [Elfes, 1989] associated with places and defines an algorithm to match two places. This method is applied to spatial learning of a dynamic indoor environment and re-localization in dynamic indoor environments [Yamauchi and Langley, 1996]. On the contrary, a statistical system attempts to describe the observations coming from the sensors as a random process that must be modeled effectively. The recognition process is envisioned as the association of the signal acquired from sensors with a model of the place to identify. For instance, in [Rompais et al., 1995], a neural network extracts places in an indoor environment from the data given by a distance measurement system based on a panoramic laser telemeter and uses it to recognize the places previously learned.

These two approaches are complementary. In the first approach, we try to understand the observations and build a representation of the observations, whereas in the second approach we build models that represent the statistical properties of the observations.

Stochastic modeling is a flexible method for handling the large variability of complex temporal signals. In contrast to dynamic time warping where heuristic training methods for estimating templates are used, stochastic modeling allows probabilistic and automatic training for estimating models.

In robotics, [Simmons and Koenig, 1995] used stochastic models based on Partially Observable Markov Decision Process (POMDP) for developping navigation methods. Their objective is to find, given an observation, the best action to reach a predefined goal, whereas our aim is to find, given an observation, the place recognized. Moreover, they do not use Markov models for sensor interpretation and include actions of the robot in their model. On the contrary, in our work, we use Markov models for sensor interpretation, but planning and execution of actions are performed by other techniques [Morignot et al., 1997].

In this section, we present a method for learning and recognizing places based on second-order hidden Markov models. HMM2 have been shown to be efficient models for capturing temporal variations in various signals and in many cases they surpass first-order hidden Markov models when the trajectory in the state space has to be accounted for. We use them to learn and recognize places by a mobile robot running in an indoor environment.

1.2   DESCRIPTION OF OUR ROBOT

Our robot (Figure 5.1) is a Nomad200 commercialized by Nomadics [Nomadics, 1996]. It is composed of a base and a turret. The base is formed by 3 wheels and tactile sensors. The turret is a uniform 16-sided polygon. On each side, there is an infrared and an ultrasonic sensor. The turret can rotate independently of the base.

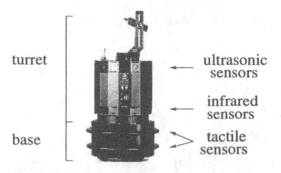

Figure 5.1. The Nomad200 robot.

1.3   TACTILE SENSORS

A ring of 20 tactile sensors surrounds the base. They detect contact with objects. They are only used for the emergency cases. They are associated with low-level reflexes as emergency stop and backward movement.

1.4   ULTRASONIC SENSORS

The angle between two ultrasonic sensors is 22.5 degrees, and each ultrasonic sensor has a beam width of approximately 23.6 degrees. By examining all 16 sensors, we can obtain a 360 degree panoramic view fairly rapidly. The ultrasonic sensors give range information from 17 to 255 inches. But the quality of the range information greatly depends on the surface of reflection and the angle of incidence between the ultrasonic sensor and the object. We have to deal with several erroneous reflections as shown in Figure 5.2.

1.5   INFRARED SENSORS

The infrared sensors measure the light differences between emitted light and reflected light. They are very sensitive to the ambient light, the object color and the object orientation. As we assume that for short distances, the range information is acceptable, we only use infrared sensors for the areas shorter than 17 inches, where the ultrasonic sensors are not usable.

1.6   ODOMETRY MEASUREMENTS

Odometry measurements integrate the translation and rotation of the robot and updates the position and orientation of the robot. As with all odometric systems, it accumulates errors during movements. We use it to have a rough idea of the position and orientation of the robot.

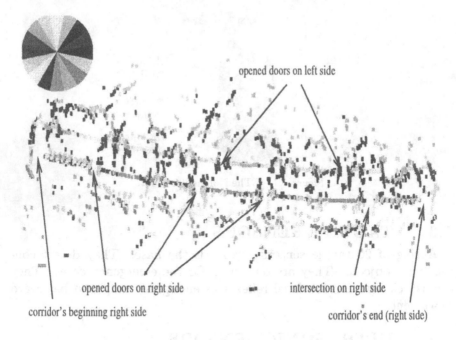

Figure 5.2. What the robot listens for when navigating in a corridor using its ultrasonic sensors (see [Aycard et al., 1998, Aycard, 1998]). Note the high density of erroneous echoes.

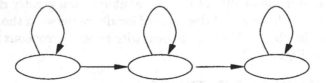

Figure 5.3. Topology of states used for each model of place

1.7   PLACE MODELING USING HMM2

In this formalism, each place to be recognized is modeled by an HMM2 whose topology is depicted in Figure 5.3. We have chosen to model five distinctive places that are representative of our office environment: a corridor, a T-intersection, a "starting" angle when the robot moves away from the angle, and an "ending" angle when the robot arrives at this angle and an open door (Figure 5.4). This set of items is a complete description of what the mobile robot can see during its run. All other unforeseen objects, like people wandering along in a corridor, are treated as noise.

In this experiment, we have to handle several major tasks: designing efficient algorithms for training and recognition purposes; collecting a corpus of observations during several runs and labeling this corpus by finding the temporal borders of each item that the robot has observed during its run.

Recognition is carried out by the Viterbi algorithm [Forney, 1973] which determines the most likely state sequence given a sequence of observations. The learning of the models is done using the maximum likelihood estimation criteria that determines the best parameters of each model given the corpus of items. It must be noted that this criteria does not try to separate models like a neural network does, but only tries to increase the probability that a given model generates its corpus independently of what the other models can do.

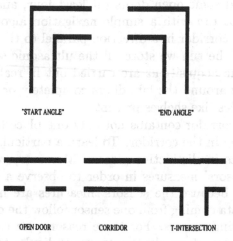

Figure 5.4.   The places to learn

The robot's environment is described by means of a grammar that enables some sequences of models and restrict others. According to this grammar, all the HMM2s are merged in a bigger HMM on which the Viterbi algorithm is used. The best sequence of states determines the ordered list of places that the robot saw during its run. It must be noted that the list of models is known when the run is completed.

1.8   APPLICATION TO MOBILE ROBOTICS
1.8.1   CORPUS COLLECTING AND LABELING

We built a corpus to train a model for each of the 5 places. For this, our mobile robot makes 50 passes (back and forth) in a very long corridor (approximatively 30 meters). This corridor (Figure 5.5) con-

Figure 5.5.   The corridor used to make our learning corpus

tains two angles (one at the start of the corridor and one at the end), a
T-intersection and some open doors (at least four, and not always the
same). The robot ran with a simple navigation algorithm to stay in
the middle of the corridor in a direction parallel to the two walls of the
corridor. During the run we store all the ultrasonic sensors' measures
of our robot. The acquisitions are carried out in real conditions with
people wandering around the lab, doors completely or partially opened
and static obstacles like shelves present.

A pass in the corridor contains not only one place but all the places
seen while running in the corridor. To learn a particular place, we need
to segment passes in distinctive places. Moreover, we need to select
the pertinent sensors' measures in order to observe a place. This task
is more complex because the sensors' measures are noisy. Figure 5.6
shows how the data coming from one sensor follow the data coming from
the two neighboring sensors. For these reasons, we choose to segment
using the data coming from the sensor perpendicular to each wall of the
corridor and its two neighboring sensors. These three sensors normally
give valid measures. As all places except the corridor cause a noticeable
variation on these three sensors over time, we define the beginning of a
place when the first sensor's measure suddenly increases and the end of
a place when the last sensor's measure suddenly decreases. Figure 5.6
shows an example of the segmentation on the right side with these three
sensors of a part of an acquisition corresponding to a T-intersection.
The first line segment is the beginning of the T-intersection (sudden
increase on the first sensor), and the second line segment is the end of
the T-intersection (sudden decrease on the third sensor). The left part
of the first line and the right part of the second line are a corridor place.
Figure 5.7 shows the position of the robot at the beginning and at the
end of the T-intersection and the measures of the three sensors used at
these two positions for the segmentation.

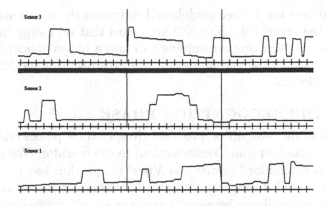

Figure 5.6. The segmentation corresponding to a T-intersection

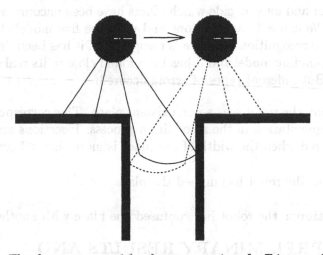

Figure 5.7. The three sonars used for the segmentation of a T-intersection

1.8.2 MODELS TRAINING

As said earlier, we have chosen three coefficients corresponding to the three sensors' measures. Because the segmentation was done using the variations on these three sensors' measures, we use the first derivative of the three sensors' measures. The topology used to train each model is shown in Figure 5.3. Intuitively, we think that the first state will contain the strong increase of the signals corresponding to the beginning of the place, the second state will contain the stationary part of the signals (where the derivative is nearly equal to zero) and the third state will contain the end of the place where the signal decreases strongly.

Two different kind of training are carried out. The first uses segmented data and each model is trained independently on these data.

The second uses the former models and estimates them on unsegmented data as in the recognition phase. This means that we merge the models seen by the robot during a complete run into a bigger model according to the sequence of observed items and train the resulting model with the unsegmented data.

1.8.3   THE RECOGNITION PHASE

The goal of the recognition process is to spot the 4 places (start-angle, end-angle, open-door and T-intersection) in the corridor. We use a fifth model for the corridor because the Viterbi algorithm has to map each frame to a model during recognition. The corridor model connects 2 items much like a silence between 2 words in speech recognition. During this experiment, the robot uses its own reactive algorithm to navigate in the corridor and must decide which places have been encountered during the run. We chose 40 acquisitions and used the five models trained to perform the recognition. A place is recognized if it has been detected by the corresponding model and it has been found close to its real geometric position. But different types of errors occured:

Insertions: the robot sees a non existent place. This corresponds to an over-segmentation in the recognition process. Insertions are actually considered when the width of the place is more than 80 centimeters;

Deletions: the robot has missed the place;

Substitutions: the robot has confused one place with another.

1.9   PRELIMINARY RESULTS AND DISCUSSION

Start angle place and End angle place are very well recognized. These places have a very particular pattern, and can hardly be confused with one another.

T-intersections are sometimes confused with open doors and open doors with T-intersections. We should merge these two places for a fair assessment because the signal provided by the sensors does not carry the information for discriminating theses two models. The corridors are roughly as narrow as the doors. A level incorporating knowledge about the environment should fix this problem.

Most of the insertions are due to the inaccuracy of the navigation algorithm and to unexpected obstacles. Sometimes the mobile robot has to avoid people or obstacles, and in this case it does not always run parallel to the two walls and in the middle of the corridor. These conditions cause reflections on the three sensors which are interpreted as places.

1.10 IMPROVEMENT OF THE RECOGNITION RATE

To improve the recognition rate, we have to solve two major problems. First, wee have to reduce the confusion rate between open-doors and T-intersections. For this, we have to find a way to improve the discriminating power of these two HMM2s. These two types of places have similar patterns but different widths. An open door has a width of 90 centimeters, wheareas a T-intersection has a width of 120 centimeters.

So far, observations have been produced on a time basis. Because the robot has a variable speed depending on its local environment, the distance travelled between two observations is not constant. Our idea is to transform these temporal observations into spatial information by taking into account the robot's speed. This will help HMM2s to discriminate between T-intersections and open doors using a width criteria.

The second issue is the excessive insertion rate. At the present time, learning and recognition of places is performed on each side independently using only the three lateral sensors (*i.e.*, six sensors in all). This technique has two drawbacks:

- we lose information about the environment using only 2 sets of 3 sensors. Our mobile robot has 16 sensors. Information provided by front and rear sensors are not currently taken into account. We have insertions of start or end of corridor in the middle of the corridor, which makes no sense. Front sensors could avoid these insertions.

- Learning and recognition are performed on each side independently. This does not permit learning and recognition of places in a global way. More global learning and recognition could be useful for several reasons. For example, the influence of one side's recognition on the other could be taken into account, to eliminate some obvious insertions.

Our idea is to process the overall observations of 16 sensors. The next two subsections present the two major modifications we used to improve the recognition rate. In the third subsection, we discuss the new results.

1.10.1 TRANSFORMATION OF OBSERVATIONS

To easily discriminate between the open-door place and T-intersection place by their widths, it's necessary that the distance travelled by the robot between two acquisitions remains constant. Our mobile robot, unfortunately, is not capable of sampling the observations as a function of the travelled distance, so we apply a variable time-sampling rate depending on the robot's speed in order to emulate a spatial sampling rate in which the distance between two acquisitions is constant.

We analyzed the travelled distance between two acquisitions assuming that the error of the odometric position estimate is negligible during short durations. We noticed that:

- in a cluttered area, the robot moves approximatively 1 centimeter between two acquisitions;

- in a clear area, the robot moves approximatively 5 centimeters between two acquisitions.

We decided that the robot has to sample an acquisition each 7.5 centimeters. We voluntary chose a longer distance than 5 centimeters to have enough data to reliably interpolate a spatial observation. We estimated a spatial sample by averaging all the data coming from each sensor during the time it takes for the robot to take an elementary 7.5 centimeter step.

1.10.2   USING 16 SENSORS FOR THE LEARNING AND RECOGNITION PHASE

An HMM2 now models the 16 dimensional real-valued signal coming from the battery of ultrasonic sensors. We next have to define more global places, taking into account what can be seen on the right side and on the left side simultaneously.

To build the new global places, we combine the 5 previous places observable on the right side with the 5 places observable on the left side. These 25 models can be broken down into the following 10 models as schown in Figure 5.8:

- a corridor;

- a T-intersection on the right (resp. left);

- an open door on the right (resp. left);

- the start of a corridor on the right (resp. left);

- the end of a corridor on the right (resp. left);

- two open doors across from each other.

1.10.3   NEW RESULTS AND DISCUSSION

Training and recognition are done once with the first derivative of the 16 sensors' measures as inputs of each HMM2.

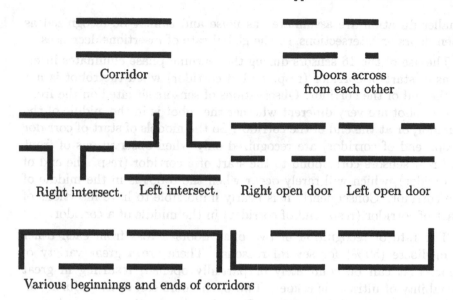

Figure 5.8.   The 10 models to recognize.

We label the same previous learning corpus using the classes defined in Section 1.10.2. We perform the learning phase of the 10 new places using the new segmentation. The recognition phase is performed using the same previous recognition corpus (Section 1.8.3). The goal of the recognition process is to spot the 9 places in the corridor between its start and its end.

We notice that the rate of confusion between open doors (on the left or on the right) and T-intersections (on the left or on the right) decreases when we use spatial samples. Due to the transformation of observations, the number of observations labeled as an open door (resp. T-intersection) is nearly always the same. As the number of observations for open doors is different from the number of observations for intersections, the discrimination between these two types of places is easier thanks to the duration modeling capabilities of HMM2 as shown in [Mari et al., 1997].

Without the transformation we had a significant number of insertions of T-intersections, and zero insertion of open doors. With the transformation we have no more insertions of T-intersections, but a higher number of insertions of open doors. In fact, a T-intersection is wider than an open door, so narrow insertions corresponding to a reflection are rarely recognized as T-intersections, but rather as open doors. Reflections of smaller duration are assimilated as noise and cannot be recognized as open doors or intersections, so the global rate of insertions decreases.

The use of the 16 sensors during the learning phase eliminates insertions of start of corridor (resp. end of corridor) when the robot is not at the end of the corridor. Observations of sensors situated on the front of the robot are very different whether the robot is in the middle of the corridor, or at the end of the corridor. So the models of start of corridor (resp. end of corridor) are recognized only when observations of front and rear sensors correspond to the start of a corridor (resp. the end of a corridor), which will rarely occur when the robot is in the middle of the corridor. Consequently it is nearly impossible to have insertions of start of corridor (resp. end of corridor) in the middle of a corridor.

The rate of recognition of two open doors across from each other is mediocre (50 %) for several reasons. There are a great variety of doors that can be completely or partially opened, resulting in great variability of ultrasonic echoes. Two open doors can overlap during an unknown number of observations. Each door is a particular case. A good investigation is to use a general *a priori* model and to adapt it depending on the position of the door. *a priori* estimation can tackle the problem of the lack of the training corpus inherent to the great number of models.

Finally, the global rate of recognition is increases from 75 % to 92 %. Insertions of places decreases from 54 % to 42 %. Omissions remain at a very low probability level (less than 1.5%).

1.11   PERSPECTIVES

In this section, we have presented a new method for learning and recognizing places in an indoor environment with second-order hidden Markov models. One of the main interests of this approach is the specification of an automatic learning algorithm of the environment that allows the recognition of typical places. The transformation of observations to simulate a spatial sampling of observations improves the global rate of recognition by decreasing the confusion between open doors and T-intersections. The use of 16 sensors instead of 2 sets of 3 allows the environment to be learned and recognized in a more general way which decreases the rate of insertions by eliminating insertions of start or end of corridor in the middle of a corridor. This method gives good results and presents a good robustness to noise. The results can be improved by adding more models to decrease the intra-class variability (especially for open doors across from each other) and to take into account some contextual information. This method has two drawbacks: the list of places is known when the run is completed and a place is recognized when it has been completely visited. For example the robot has to go back to turn at a T-intersection after it has recognized it.

2. APPLICATION OF MARKOV DECISION PROCESSES

Markov decision processes (MDPs) have been successfully used in robotics for indoor robot navigation problems. They allow to compute optimal sequences of actions in order to achieve a given goal, accounting for actuators uncertainties. But MDPs are weak to avoid unknown obstacles. At the opposite reactive navigators are particulary adapted to that, and don't need any prior knowledge about the environment. But they are unable to plan the set of actions that will permit the realization of a given mission. We present a state aggregation technique for Markov decision processes, such that part of the work usually dedicated to the planner is achieved by a reactive navigator. Thus some characteristics of the environments, such as width of corridors, have not to be considered, which allows to cluster states together, significantly reducing the state space. As a consequence, policies are computed faster and are shown to be at least as efficient as optimal ones. The results presented in this section are taken from [Laroche et al., 1999]

2.1 INTRODUCTION

Applying Markov decision processes [Bellman, 1957] to robotics has been the subject of several studies [Dean et al., , Cassandra et al., , Koenig and Simmons, 1998]. Compared to the traditional robot navigation approaches, MDPs allow to explicitly represent the various form of uncertainties present in navigation problems: actuator and sensor uncertainties, dynamic environments, uncertainty about the initial position of the robot, etc. Using a MDP policy, all these uncertainties are considered, generating plans with high success probability. But classical algorithms used to compute these policies are intractable for solving large MDPs requiring a large state space. As a consequence, aggregation [Boutilier and Dearden, 1994, Laroche and Charpillet, 1998] and decomposition [Parr, 1998, Hauskrecht et al., 1998, Precup and Sutton, 1997] techniques have been the subject of a lot of recent studies, but only a few of them have been applied to robotics.

Among studies applying MDPs to robotics, [Dean et al.,] uses the knowledge of the initial position of the robot to consider only a subset of the state space to compute the policy. This subset of states is designed using an *envelope* containing states on a path between the initial and goal states. Once a policy has been computed on this first envelope, it is sent to the robot which executes this policy, while the planning algorithm iteratively extend the initial envelope to compute policies considering more states. Each time a policy has been computed, it is sent to the robot; the process iterates until the robot has reached its goal or all states are contained in the envelope. The problem with this approach is that as not all states are considered during first policies computing, these ones can be very sub-optimal. The time gained using this method could be outweighed by turn-backs of the robot when two successive policies define conflicting actions. Furthermore, this study only consider actuator uncertainties, and assumes perfect sensor information that allow to localize the robot perfectly, which is not realistic. In contrast, [Koenig and Simmons, 1998] has a very realistic approach of robotics problems, using a partially observable MDP to cope with uncertainties at execution time. This work is nevertheless quite different from other approaches, as it only uses stochastic processes to localize the robot, but not to compute policies. [Cassandra et al.,] also uses MDPs in a robotics application, and gives very interesting results on various MDP policies execution strategies. But as no aggregation technique is applied to their model, its utilisation is reduced to quite small problems.

The main originality of the approach presented here is to show how robotics domain specificities can be used to aggregate states without loss of optimality. A MDP policy defines actions to avoid obstacles, and generates behaviours taking away the robot from walls of corridors. The idea is that as these behaviours can easily be generated by a navigator, the MDP policy has not to cope with these problems. So usually important characteristics of the environment, such as width of corridors, have not to be considered explicitly. This is the basic idea of the aggregation technique, which allows to considerably reduce both state space and computing time of policies. The approximate policies presented here are shown to be at least as efficient as optimal ones, as shown on 500 experiments made with the robot. In the following section, we recall the MDP formalism (see also (4.6) for an introduction). Section 2.3 presents the architecture we use to execute the stochastic policies, and the way states are clustered is explained in Section 2.4. Results are given in Section 2.5, we compare the approach presented here with previous work in section 2.6.

2.2   MARKOV DECISION PROCESSES

Recall that a Markov decision process (cf. page 18) models an agent which interacts with its environment, taking as input the states of the environment and generating actions as outputs. It can be formally defined as follows (see):

- $\mathcal{S}$ is a finite set of states of the environment;

- $\mathcal{A}$ is a finite set of actions;

- $T : \mathcal{S} \times \mathcal{A} \times \mathcal{S} \rightarrow [0, 1]$ is the state transition function, which encodes the probabilistic effects of actions. $T(s, a, s')$ is the probability of moving to state s' when action a is executed in state s.

- R is a reward function used to design the goal the agent has to reach and dangerous parts of the environment, such as stairwells. According to the problem to solve, it can be quite simple ($R : \mathcal{S} \rightarrow \mathbb{R}$; $R(s)$ gives the reward the agent gets for being in state s) or more complex ($R : \mathcal{S} \times \mathcal{A} \times \mathcal{S} \rightarrow \mathbb{R}$; $R(s, a, s')$ gives the reward the agent gets starting in state s, choosing action a and reaching s').

In this framework, it is assumed that effects of actions may be uncertain, but once the action has been executed, the agent has perfect perceptual abilities accurately determining the current state. This assumption is not realistic, specially for a mobile robot, but we will deal with that at execution time, as we will see later.

2.2.1   PLANNING ALGORITHM

Once the environment has been modelled and the goal chosen, an optimal policy π defines for each state of $\mathcal{S}$ an optimal action to be executed. Several optimality criteria can be used; here we will focus on *infinite* horizon *discounted* decision problems. The optimal policy maximizes the value of each state which is defined by the expected rewards the agent may obtain. Two main algorithms are usually used to compute such an optimal policy for an MDP, `Value Iteration` [Bellman, 1957] and `Policy Iteration` [Howard, 1960]. After some empirical studies, we choose the second one, as it better fits with the model.

`Policy Iteration` iteratively maximizes the value function, which is formulated as follows:

$$V_\pi(s) = R(s) + \gamma \sum_{s' \in \mathcal{S}} T(s, \pi(s), s') V_\pi(s') \qquad (5.1)$$

γ is a discounting factor used to give more or less importance to future rewards. To find a policy, a set of $|\mathcal{S}|$ linear equations in $|\mathcal{S}|$ unknowns has to be solved. The initial policy is successively improved until an optimal one is found:

1. $\pi' =$ any plan

2. While $\pi \neq \pi'$

   a. $\pi = \pi'$

   b. For all $s \in S$

      Compute $V_\pi(s)$ by solving the system of $|S|$ equations in $|S|$ unknowns given by equation 5.1

   c. For all $s \in S$

      If there exists an action $a \in A$ such that:

      $R(s,a) + \gamma \sum_{s' \in S} T(s,a,s') V_\pi(s') > V_\pi(s)$

      Then $\pi'(s) = a$

      Else $\pi'(s) = \pi(s)$

3. Return π

Complexity results for MDPs can be found in [Papadimitriou and Tsitsiklis, 1987] and [Littman et al.,]. Each iteration of the algorithm consists of two steps: solving the linear equation system, which can be done in $O(|S|^3)$ operations, and the improvement of the current policy, which needs $O(|A||S|^2)$ operations. The algorithm is guaranteed to converge [Howard, 1960], and generally tends to do so in a few number of iterations [Puterman, 1994].

2.2.2   MODEL USED

As stated in the previous section, an MDP is defined using several parameters. The transition and reward functions are central in the construction of the optimal policy. The way obstacles of the environment are represented is also very influent on the optimal actions computed for states near these obstacles.

Details on the model can be found in [Laroche and Charpillet, 1998]. The environment is represented using a grid of 20-inch squares; in each square, the robot can be oriented along four directions (North, South, East and West), thus four states are necessary to represent one square of the grid. Three actions have been chosen: go forward to the next square, quarter turn on the left and quarter turn on the right. We use a discounting factor $\gamma = 0.99$, which appeared to be well adapted to the problems. The reward function R is defined in a very simple way: if the robot is in the goal state, it gets a reward of 0; it gets -1 in all other states. The goal is an absorbing state (once the robot is in this state, it can not go out of it), just like the obstacles. That makes the value of a state s_o containing an obstacle highly negative:

$$V_\pi(s_o) = R(s_o) + \gamma(1 \times V_\pi(s_o)) \rightarrow V_\pi(s_o) = \frac{R(s_o)}{1 - \gamma}$$

With the reward function and the γ factor used, this leads to a value of -100 which is the worst possible value. We use a learned transition function: a Go_Forward action succeeds 57% of time, while 43% of the time the robot moves to one of ten other possible adjacent states; Turn_Left and Turn_Right actions succeed 85% of time.

2.3   MDP-BASED NAVIGATION ARCHITECTURE

The architecture we use to execute the MDP policy is depicted in Figure 5.9. It is mainly composed of three parts: the policy to execute, a state estimator to localize the robot and an action execution module. The state estimator sends the current state to the MDP policy which sends the optimal action corresponding to this state to the action execution module. The current action is then executed by a navigator, which is used to avoid unknown obstacles. When the distance covered by the robot exceeds a predefined threshold, the values of the sensors of the robot are send to the state estimator. The process then iterates until the goal is reached or a collision has occurred. In the following sections, we describe in details the state estimator and the action execution module.

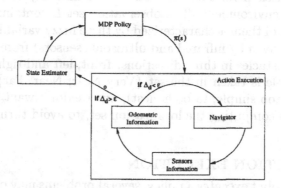

Figure 5.9.   MDP-based navigation architecture.

2.3.1   STATE ESTIMATOR

As pointed out previously, the Markov Decision Process framework assumes that states of the model are accurately identified by the agent. This implies that the robot should have perfect sensors, which is unfortunately not realistic. To solve this problem, [Cassandra et al.,] propose to use an observation function to help the robot in its localization process.

This function is defined as follows:

$$O : \mathcal{S} \times \mathcal{B} \to [0,1]$$

where $\mathcal{B}$ is a set of abstract features of the environment the robot can observe. $O(s,o)$ is the probability to make the observation o when the robot is in state s. During the execution of the policy, this function allows to define a set of possible states in which the robot may be. This will be done by the state estimator, which computes a localization set l (set of all possible current states s'), using as input the last action, the last observation and the previous localization set l_{prev}:

$$l(s') = \frac{O(s',o) \sum_{s \in \mathcal{S}} l_{\text{prev}}(s) T(s,a,s')}{\alpha}$$

α is a normalization factor that ensures that the resulting probabilities sum to 1. To decide which action to execute, we choose the one associated to the most likely state in l (*i.e.* the one with the highest probability).

The quality of the state estimator is highly dependent of the transition and observation function. To accurately reflect the environment, this function should be learned, just like the transition function. In the system presented here, this function is automatically computed using a map of the environment. The observation set $\mathcal{B}$ contains 27 possible features: each of them is characterized by three fuzzy variables indicating the distance (given by infrared and ultrasonic sensors) from the robot to the nearest obstacles in three directions: front, left and right. The value of each variable is taken in the set {Very Near, Near, Far}. It is quite simple (a bit too simple, to be honest), but we don't want to spend too much time on computing the localization set, to avoid turn-backs of the robot.

2.3.2   ACTION EXECUTION

When the robot executes a policy, several problems may occur. Firstly, there can be unknown obstacles in the environment, which may cause collisions. Secondly, the robot may be lost in the environment, and thus may execute actions which could also cause collisions. For these two reasons, we have to use a navigator to execute the actions defined by the MDP policy. The one we use [Aycard et al., 1997b] is based on fuzzy rules. The values of adjacent ultrasonics and infrared sensors are grouped on the front and the right and left sides of the robot to define a global behaviour allowing the robot to safely navigate in an unknown environment. Thanks to this navigator, the robot always tries to place itself in the middle of corridors. We use this navigator only for Go_Forward actions, as the robot is stopped before turning actions. In Figure 5.10 is shown a mission of the robot. One can see that the robot avoids obstacles in the first vertical corridor, without stopping and without turning actions. Similarly, it goes to the goal through the horizontal corridor, softly placing itself in the middle of the corridor, to avoid collisions with the walls.

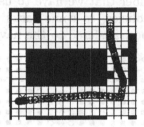

Figure 5.10. The robot navigating to the goal.

2.4   STATE AGGREGATION

2.4.1   INTRODUCTION

As classical methods are intractable for solving MDPs requiring a large number of states, methods which aim to reduce the state space are very useful to cope with large problems. Some approaches only consider a small set of states to compute the optimal policy, like the one of [Dean et al.,] we expose in the introduction of this paper. Other techniques consider all states to construct approximate policies, but the state space is either *decomposed* or *aggregated*. Decomposition approaches [Dean and Lin, 1995, Precup and Sutton, 1997, Parr, 1998, Hauskrecht et al., 1998] divide the global MDP into smaller MDPs which are independently solved. These local solutions are then combined to form an approximate solution for the global MDP. These methods can be very efficient, but are only usable when the number of states connecting two subproblems is small. The second class of approaches [Boutilier and Dearden, 1994, Laroche and Charpillet, 1998] clusters states together to solve a smaller problem that is an approximation of the global one. Each cluster is assigned a single action, but the reward and transition functions are transformed to take in account each *base* state of the clusters. The problem with these techniques is that it is not always easy to define clusters. Within a cluster, states have to share different characteristics to ensure "good" quality of the policy obtained. In the following section, we show on an indoor navigation problem how we use domain characteristics to cluster states together.

2.4.2   THE METHOD

Aggregation and decomposition techniques described previously aim to compute a "good" approximate policy, reducing computing time. The implicit assumption is that a good approximate policy will be nearly as efficient as an optimal one. As a contrast, the aggregation technique presented here gives a very *bad* policy which is nevertheless very efficient. The key idea is to remember that we have to use a navigator to ensure the good execution of policies, so some tasks usually devoted to the planning algorithm can be transfered to the navigator.

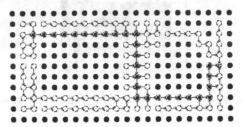

Figure 5.11.   An optimal policy.

To explain the method, we use the little example depicted by Figure 5.11. This figure shows an optimal policy computed for a 440-states environment. The goal to reach, in the right vertical corridor, is labelled with a 'G'. Each position of the environment is symbolized by a circle (filled if it is an obstacle). For each state, the optimal action is represented by an arrow between two circles (action Go_Forward) or on the circle (actions Turn_Right and Turn_Left, as there are four possible orientations for each position of the environment). The partially filled circles show the optimal path between a top left state and the goal state. This example shows clearly two specificities of an MDP policy:

1. Each translation of the robot may lead to a collision with a wall, so states adjacent to walls have a higher cost than states in the center of corridors. As a consequence, the policy moves away the robot from the walls. This can be verified in the central vertical corridor of the environment, in which each state adjacent to a wall aims the robot to central states of the corridor.

2. The optimal paths are not necessarily the shortest ones. Following the optimal path shown by the partially filled circles on Figure 5.11, the robot will reach the goal after having executed (at least) 31 actions. A shortest way exists: just going through the two top horizontal corridors before turning to the right would lead the robot to its goal after only 25 actions. But the top right horizontal corridor is cluttered

with several obstacles. Thus states in this corridor have a high cost, which explains why the optimal path goes along a longer way.

In a robotics application, this second characteristic is very interesting. The 25-actions path is shorter than the optimal one, but it is in fact slower and harder to execute. Navigating through such a cluttered corridor could be dangerous for the robot. Even if it achieves to accurately navigate through these obstacles, its speed has to be very slow in this kind of situation. So even if the path is shorter, it is in fact slower. The robot reaches its goal more safely and more quickly using the optimal path. As a contrast, the first characteristic (moving away from walls and obstacles) is less interesting. It is a very simple task for the navigator to aim the robot to the center of corridors. Whatever the width of a corridor (or an office), the navigator is sufficient to achieve this task. Moreover, the navigator can easily avoid obstacles, as it is shown in Figure 5.10. So the idea is to build an abstract MDP which generates paths avoiding cluttered corridor, but without loosing time specifying actions to move away the robot from walls or obstacles.

a. Without state aggregation : 212 states.   b. With state aggregation : 30 states.

Figure 5.12.   State aggregation in a corridor.

To achieve that, we build an abstract MDP, aggregating states in corridors (we can also apply this to offices, but in the following we will only consider corridors). To distinguish states between corridors, offices and intersections, we use a map of the environment in which each state is labelled. The aggregation technique is depicted in Figure 5.12. This figure shows a 15-states long 4-states large corridor. In a classical MDP, this corridor would be represented using 212 states ($15 \times 4 \times 4 = 240$ minus the positions taken up by obstacles). As the navigator can naturally lead the robot to the center of corridors, their width is not important. Thus we can aggregate the four positions in a single cluster. This leads to a corridor which would be represented by 60 states (15 positions with 4 orientations). When the robot is in an intersection of the environment (or in front of an office door), four orientations per state are needed to possibly change direction. But in a corridor, some orientations are unnecessary. In a horizontal corridor, the North and South orientations have no importance: what the robot has to do is to go through the corridor from east to west or vice versa. So we just need two orientations for positions in corridors: East and West for a horizontal corridor, North and South for a vertical one. The corridor is now only represented by 30 states, which is 1/7 of the state space of the classical MDP. As we keep only two orientations in corridors, the Turn_Left and Turn_Right actions are not usable for these states. So two new actions are defined: Cross_Corridor and Turn_About. They are only available in the abstract states (*i.e.* states in corridors). This also contributes to simplify the MDP, as only 2 actions are possible for these states, whereas there were 3 possible actions for every state in the classical MDP.

Another problem to solve is to transform the transition function to encode the new actions and the new state space. The learning phase gave us a transition function usable for the 20 inches large states, which will not be usable for the new model. To build the Cross_Corridor part ofthe new transition function, we consider all the Go_Forward transitions for the states s composing the corridor c:

$$T(c, \text{Cross\_Cor}, s') = \frac{\sum\limits_{s \in c} T(s, \text{Go\_Forward}, s')}{N}$$

where N is the number of states in the corridor, used to ensure that the probabilities will sum to 1. The transitions for the Turn_About action are computed applying two successive turning actions: starting in a corridor and executing this action results in a self-transition with probability 0.14; the action is successfully executed with probability 0.86. The reward function has not to be changed, each abstract state has a -1 reward, just like base states.

To sum up, we have now two types of states in the abstract MDP. Firstly, there are classical states which are states composing intersections or states placed in front of an office door. In these states, the robot has to be able to turn to change direction. As a consequence, these states are not aggregated: no cluster is created, and four orientations are kept. In these states, three actions may be executed: Go_Forward, Turn_Right and Turn_Left. Secondly, there are abstract states which are states placed in corridors or offices, which contain several base state. In these states, only two orientations and two actions are kept: Cross_Corridor and Turn_About.

This gives an aggregation method which is very fast. Once the aggregation is completed, the abstract MDP is solved using Policy Iteration. The approximate policy π_a can theoretically be used as is, but to accurately execute it, we have to transfer it to the classical MDP containing all states. This is done using Algorithm 5.1, which transforms a policy

π_a of an abstract MDP $\mathcal{M}_a$ into a policy π_c of a classical MDP $\mathcal{M}_c$. Basically, the optimal action of each state of $\mathcal{M}_a$ is transformed into an action for the corresponding state in $\mathcal{M}_c$. For classical states, no transformation is needed: the action for the state of $\mathcal{M}_c$ is the one defined for the state of $\mathcal{M}_a$. For an abstract state, its action is affected to every *base* state contained in the cluster. If the base state has the same orientation as the abstract one, the action is simply affected to the base state; but if the base state s' has an orthogonal orientation, a turning action is affected to it. A Cross_Corridor action is simply changed into a Go_Forward action; Turn_About becomes two successive Turn_Right actions.

Algorithm 5.1 Transforms a policy π_a of an abstract MDP into a policy π_c of a classical MDP

> **for all** $s \in \mathcal{S}_{\mathcal{M}_a}$ **do**
>> **if** is_an_abstract_state(s) **then**
>>> **for all** $s' \in s$ **do**
>>>> **if** same_orientation(s, s') **then**
>>>>> $\pi_c(s') = \pi_a(s)$
>>>> **else**
>>>>> $\pi_c(s') = $ Turn_to_reach(s, s')
>>>> **end if**
>>> **end for**
>> **else**
>>> $\pi_c(s) = \pi_a(s)$
>> **end if**
> **end for**

In the next section we present some results obtained using the aggregation method. Here we briefly show that the navigator can be used to safely execute approximate policies. In Figure 5.13 is shown the approximate policy computed using the method for the sample environment. This MDP contains only 204 states, whereas the classical MDP needs 440 states. Two main differences can be found in comparison with the optimal policy (Figure 5.11). Firstly, no effort is made to avoid going along walls: in the central vertical corridor, this policy is very different from the optimal one. Secondly, the approximate policy does not avoid obstacles in the top right corridor. In a lot of not too complex situations, a navigator should be able to safely avoid obstacles. In spite of these differences, the optimal path leading the robot from the top left state to the goal is the same according to both policies. So we obtain a policy which is very far from the optimal one (as not avoiding obstacles and going along walls has a cost), but which have a similar global behaviour.

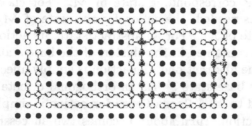

Figure 5.13.   The policy obtained with the method.

To illustrate this, Figure 5.14 shows two missions executed by the robot, once following the optimal policy, and once following the approximate policy. The first mission (Figure 5.14(a,b)) shows clearly that the behaviour of the robot is the same following the two policies, specially in the central corridor. As the navigator naturally places the robot in the central part of the corridor, the actions specified by the optimal policy to move away the robot from the walls are not used. The second mission (Figure 5.14(c,d)) starts close to a wall of the central corridor. Following the optimal policy, the robot immediately reaches the middle of the corridor. It is then stopped, before turning to its right, and following its path to the goal. At the opposite, the approximate policy leads to a more natural behaviour: the navigator softly places the robot in the central part of the corridor, without any turning actions.

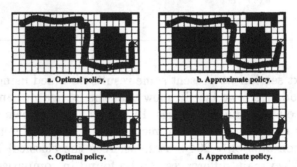

Figure 5.14.   2 missions executed using (a,c) the optimal policy ; (b,d) an abstract policy.

2.5   RESULTS

The method has been tested on several environments, from small, 440-states environments (Figure 5.11) to bigger ones with more than 10000

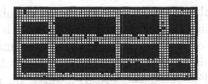

Figure 5.15. Test environment (3976 states).

states. Here we present results on a big environment (Figure 5.15). This is a classical office environment, in which some corridors are partially obstructed by obstacles. The goal has been successively placed in 20 different states, so results presented are an average on these 20 problems. Table 5.1 shows that approximates policies are computed very faster than optimal ones. The classical MDP is represented using more than 3900 states, whereas the abstract MDP contains only 1238 states (state space divided by 3.2). This allows to divide the computing time of a solution by nearly 3, as the optimal solution is obtained after 1690 seconds (on average), whereas the method gives a solution after 570 seconds. The time needed to build clusters of states and modifying the transition function is included in these results; it is in fact negligible. These results are very promising, but are not always as good: on smaller environments, with narrower corridors, aggregation can not reduce the state space as efficiently. However, for the class of environments we use, dividing state space (and computing time) by 2 is the worst performance we have obtained.

	Without aggregation	The method
Number of states	3976	1238
Computing time	1690	570

Table 5.1. Aggregation benefit.

To evaluate the quality of an approximate policy π_a according to the optimal one, a classical method is to use the maximum Bellman error [Williams and BairdIII, 1993]. This method is very useful as it can be used without knowing the optimal solution, but it only gives a bound on the distance to the optimal. But when the optimal policy π^* is known, a better evaluation can be computed, using the ratio r, as follows:

$$r = \frac{V_{\pi_a} - V_{\text{worst}}}{V_{\pi^*} - V_{\text{worst}}}$$

where V_{worst} is equal to the value of the worst possible policy. Another estimate of the quality of the approximate policies is given by a second value: the number of states s such that the optimal action $V_{\pi_a}(s)$ is equal to $V_{\pi^*(s)}$. This measure is given as a percentage over the total number of states. For these two criteria, three measures are given in Table 5.2: average, worst and best performances. These results are very *bad*. On the same environment, using a decomposition technique, we reached an average value of more than 96% of the optimal. But this is not surprising at all ! Remember the approximate policies we generate don't avoid obstacles and don't place the robot in the center of corridors. So states in the abstract MDPs have a higher cost, which explains these bad results. They are only given as an indication, but are in no way an estimation of the real quality of the method.

	Value	Actions
average	52.58	67.81
worst	40.21	65.67
best	68.16	70.22

Table 5.2.   Results for the test environment.

The real estimation of the approximate policies efficiency is given in Table 5.3. We have given 500 missions to the robot in the test environment, once following the optimal policy, once following the approximate one. This table presents three values: the number of mission successfully completed, an average time needed by the robot to reach the goal, and an average number of actions executed to reach the goal. The two last results are not surprising: the robot has completed its missions faster following the approximate policy. This is due to time lost by the execution controller constantly heading the robot in the middle of corridor. Following an optimal policy, a robot placed near a wall is stopped, it turns, goes to the middle of the corridor, turns and stops again, and then continues its mission. Following the approximate policy, the robot just cross corridors, without taking care of its position with respect to the walls. Thanks to that, the robot does not loose time executing unnecessary turning actions. At the opposite, the number of success, which is better using an approximate policy, is more surprising. It appeared in fact that turning actions increase the risk of mislocalization of the robot. That explains why approximate policies succeed more often.

	Optimal policy	The method policy
Number of success	390	420
Average time	311	244
Average actions	107	80

Table 5.3.   Policy Execution.

2.6   COMPARISON WITH PREVIOUS WORK

In [Laroche and Charpillet, 1998], two other aggregation techniques are proposed. At first, these could seem very similar to the present work. The basic idea was also to aggregate states in corridors, but clusters were defined less precisely: all states contained in a part of corridor between two doors were aggregated into a single abstract state. The state space was thus much more reduced, but this method was only usable for environments with not too long corridors. Defining a single action for a 20-meters long corridor may lead to very sub-optimal policies. Furthermore, these methods aimed to approximate optimal policies, but without taking into account the use of a navigator. The method we present here needs more time to compute a policy (as there are more states to consider), but generates more efficient policies. In the near future, we plan to combine the two approaches. In each corridor, we can define a big cluster containing all states, and small clusters defined as in this paper. The action defined for the big cluster can thus be used as a macro-action (as in [Parr, 1998]) usable in the whole corridor. This should accelerate the approximate policy computing.

2.7   CONCLUSION

Robotics and stochastic planning are two areas of research which have been extensively studied, but approaches aiming to use MDPs in robotics are more unusual. The main contribution of this work is to show that a theoretically bad policy can be very efficient in mobile robotics, as a navigator can achieve tasks usually encoded in the policy to execute. The aggregation technique is very simple and fast. It necessitates only a description of the environment in which corridors (and offices) are labeled. It allows to reduce the computing time of policies, giving sub-optimal but very efficient policies.

The approach presented is quite robust, but some environments cluttered by many states are still problematic. To improve results in this kind of situations we explore several solutions like combining this method with the ones we had proposed in the past. We should also improve the way we cope with perceptual capabilities of the robot, to ensure a better rate of successfully completed missions.

References

This chapter is based on the articles [Laroche et al., 1999, Aycard et al., 1997a, Aycard et al., 1998]

Appendix A
Some useful statistical programs

1. THE GAUSSIAN DENSITY CLASS

This section describes a useful class of objects that we have met in this book, *the multidimensional Gaussian pdf*. Each instance of this class is a multidimensional, full covariance, *pdf*. The dimension of the space is represented by the class variable nCoef. Due to sordid computation restrictions, nCoef must be less than the constant NCOEF defined in the file application.h. Such a *pdf* is defined by 2 arrays: the mean vector mean and the covariance vector covar. Because the covariance matrix is symmetric, we only keep the items below the diagonal that are stored in the lexicographic order of the indices. Therefore, the item in row i and column j is stored at the offset $i * (i + 1)/2 + j$:

$$cov(i, j) = cov[i * (i + 1)/2 + j] \text{ with } j \leq i.$$

Before creating any objects belonging to the class Pdfg, we have to set the 2 class variables nCoef and sizeCovar. It is done by invoking the class method Pdfg::setNCoef(v) with a whatever value less than NCOEF as argument.

1.1 THE METHODS

The file Pdfg.h contains the interface of the class Pdfg. We have the useful following methods to handle a *pdf*:

distanceMaha(x) computes $(x - mean)^t \Sigma^{-1}(x - mean)$ where x is a vector, Σ^{-1} the inverse covariance matrix stored in covar and mean the mean of the *pdf*.

probG(x) computes the likelihood of x as defined in Eq. 2.1, Page 33.

resetMC() resets the mean and covariance matrices. It is done prior an estimation process.

addWeightedSample(w, x) adds the vector x weighted by w during the estimation of a *pdf*.   mean and covar play the role of accumulators.

normMat(nSamples) normalizes the mean and covariance matrices when all the samples have been collected. It also inverts the covariance and sets the determinant.

shake(f) performs a perturbation in the mean, leaving the covariance matrix unchanged. The scaling factor f controls the range of the perturbation. f=0 means no perturbation.

The inversion of a symmetric matrix is performed by means of the Cholesky decomposition [Lebart et al., 1982]. The function `igsdp()` performs the inversion. We have been testing these functions for a long time, they should be bug free. The file `Pdfg.C` contains the associated code written in C++.

```
// file Pdfg.h
// implements the class of Gaussian Pdf
#ifndef __PDFG_H
#define __PDFG_H
#include <stdio.h>
#include <iostream.h>
#include "application.h"
class Pdfg // a full Gaussian  pdf
{
 private:
    static int nCoef, sizeCovar ;
    double *mean ;     // the mean matrix
    double *covar ;    // the  inverse covariance matrix
    double det ;       // the log of covar's  determinant
    double igsdp(double *, double *, int n, int *pNRang) ;
    double chlsk(double *, double *, int n, int *pNrang) ;

 public:
    static void setNcoef(int) ;
    double* getMean() const {return mean ;}
    double* getCovar() const {return covar ;}
    void shake(double f) ;
    double probG(const double* frm) ;
```

```
// compute the likelihood of a given frame
   void resetMC() ; //   reset Mean and Covar
   double distanceMaha (const double* frm) ;
   void addWeightedSample(double weight,
                          const double *sample) ;
   double Pdfg::normMat( double nSamples) ;
// the constructors
   Pdfg( const Pdfg&) ; // copy
   Pdfg() ; // default constructor
   ~Pdfg() ;
   friend ostream& operator << (ostream & s,
                                const Pdfg & item) ;
} ;
#endif /*!__PDFG_H*/
```

The file `Pdfg.C` contains the implementation of this class written in C++.

```
// file Pdfg.C
#include <math.h>
#include <iostream.h>
#include "Pdfg.h"
#define EPSI (1.e-20)
// index of item (i,j) in a symmetric matrix
// we only store items below the diagonal
#define RG(i,j) ((i) * (i+1)/2 + (j))
#define Q(i,j)   q[RG(i,j)]
#define W(i,j)   w[RG(i,j)]
#define R(i,j)   r[RG(i,j)]
#define WM1(i,j) wm1[RG(i,j)]

double Pdfg::probG(const double *frm)
////////////////////////////////////////
// computes the gaussian prob. of frm
{
    double  prob = det + distanceMaha(frm) ;
    prob = exp(-prob / 2.) ;
    return prob ;
}
//   2 times enrollment
double Pdfg::distanceMaha (const double *frm)
////////////////////////////////////////////////
//   compute the Mahalanobis distance.
```

```
// between frm = X and the pdf
{
    double sum1, sum2 ;
    int i, j ;
    double xm[NCOEF] ; // X - M
    int ki,kk;
//
//   compute (X - M)t [sigma -1] (X - M)
//
    ki=0;
    kk=0;
    sum1 = sum2 = 0. ;
    for (i = 0; i+1 < nCoef ; i+=2)
    {
        xm[i]=frm[i]-mean[i];
        xm[i+1]=frm[i+1]-mean[i+1];
        for(j = 0; j < i ; j++)
            sum2 += xm[i] * xm[j] * covar[ki+j] +
                xm[i+1] * xm[j]* covar[ki+j+i+1] ;
        sum2 += xm[i+1] * xm[j]* covar[ki+j+i+1] ;
        sum1 += covar[kk] * xm[i]*xm[i] +
            covar[kk+i+2] * xm[i+1] * xm[i+1];
        kk += 2 * i + 5;
        ki += 2 * i + 3;
    }
    for (; i < nCoef ; i++)
    {
        xm[i]=frm[i]-mean[i];
        for(j = 0; j < i ; j++)
            sum2 += xm[i] * xm[j]* covar[ki+j];
        sum1 += covar[kk] * xm[i] * xm[i];
        kk += i + 2 ;
        ki += i + 1 ;
    }
    return sum1 + 2. * sum2 ;
}
Pdfg::Pdfg()
/////////////
// default constructor
{
    mean = new double [nCoef] ;
    covar = new double [sizeCovar] ;
```

```
    resetMC() ;
}
Pdfg::~Pdfg()
//////////////
// destructor of class Pdf_t
{
    delete [] mean ;
    delete [] covar ;
}
void Pdfg::resetMC()
//////////////////
//    reset Mean and Covar
{
    int i ;
    for ( i = 0 ; i < nCoef ; i++)
        mean[i] = 0. ;
    for ( i = 0 ; i < sizeCovar ; i++)
        covar[i] = 0. ;
    det = 0. ;
}
Pdfg::Pdfg(const Pdfg& pdf)
//////////////////////////
//constructor copy
{
    int i ;
    mean = new double [nCoef] ;
    covar = new double [sizeCovar] ;
    for(i = 0 ; i < nCoef ; i++)
        mean[i] = pdf.mean[i] ;
    for(i = 0 ; i < sizeCovar; i++)
        covar[i] = pdf.covar[i] ;
    det = pdf.det ;
}


extern "C" float ran1(int *);
extern "C" float gasdev(int *);
static int idnum = -13 ; // for random generator


void Pdfg::shake(double f)
//////////////////////
// perturb mean according to variance by a factor f
{
```

```
    for (int i = 0 ; i < nCoef ; i++) {
        double deltaSigma =  sqrt(covar[RG(i,i)]) ;
        mean[i] *= (1. + f * ran1(&idnum) / deltaSigma) ;
    }
}
void Pdfg::addWeightedSample(double weight,
                             const double *sample)
//////////////////////////////////////////////////////////
// add a (double) weighted sample to mean and covar
{
    double *pfc = covar ;

    for(int i = 0 ; i < nCoef ;  i++)
    {
        double tempo = weight * sample[i] ;
        mean[i] += tempo ;
        for (int j = 0 ; j <= i ; j++, pfc++)
            *pfc += (tempo * sample[j]) ;
    }
}
double Pdfg::normMat( double nSamples)
    /////////////////////////////////////////
    //   normalyze mean and covar
    //   returns determinant of covar
    //   double *mean ;  de la classe
    //   double *invCovar ;
    //   double *covar ; // de la classe
    //   double nSamples ; accu
{
    double *pCovar  = covar ;
    int nRang ;
    double invCovar[NCOEF*(NCOEF+1)/2] ;
    double det1 ;

    if (nSamples == 0.) return 0. ;
    pCovar = covar ;

    for (int i = 0 ; i < nCoef ; i++)
        mean[i] /= nSamples ;

    for (int i = 0 ; i < nCoef ; i++)
        for (int j = 0 ; j <=i ; j++, pCovar++)
```

```
            *pCovar = *pCovar / nSamples -
                mean[i] * mean[j] ;

    det = det1 = igsdp(covar, invCovar, nCoef, &nRang) ;
    if (det > 0.) {
        for (int i = 0 ; i < sizeCovar ; i++)
            covar[i] = invCovar[i] ;
        det = log(det) ;
    }
    return  det1 ;
}
/* inverse a symmetric positive matrix
** w(n X n) input matrix, only the items
** below the diagonal are stored
** wm1     its inverse
** n       its order
** NRang   its rank (must be n)
** return determinant of W
**              ===
*/
double Pdfg::igsdp(double *w, double *wm1,
                   int n, int *pNRang)
{
    double q[NCOEF * (NCOEF+1) / 2] ;
    double r[NCOEF * (NCOEF+1) / 2] ;
    double detW ;
    double  s ;
    int i, j, k ;
    register double *qik, *rki, *rkj, *rij ;
    register double  *wm1ij ;
// Cholesky decomposition
    detW = chlsk(w, q, n, pNRang) ;
    if (detW <= 0)
        return detW ;

    for (j = n-1 ; j >= 0 ; j--) {
        if (Q(j,j) == 0.)
            for( k = j ; k < n ; k++)
                R(k,j) = 0. ;
// generalized inverse
        else {
            R(j,j) = 1. / Q(j,j) ;
```

```
            for (i = j+1 ; i < n ; i++) {
                s = 0. ;
                qik = &Q(i,j) ;
                rkj = &R(j,j) ;
                for (k = j ; k < i ; k++, qik++, rkj += k)
                    s += *qik * *rkj ;
// s += Q(i,k) * R(k,j) ;
                if (Q(i,i) == 0.) R(i,j) = 0 ;
                else R(i,j) = -s / Q(i,i) ;
            }
        } // Q(j,j) != 0
    } // for j

    wm1ij = wm1 ;
    rij = r ;
    for (i = 0 ; i < n ; i++)
        for (j = 0 ; j <= i ; j++) {
            s = 0. ;
            for (k = i, rki = &R(i,i), rkj = rij++
                    ; k < n ; k++, rki += k, rkj += k)
                s += *rki * *rkj ;
// s += R(k,i) * R(k,j) ;
            *wm1ij++ = s ;
// WM1(i,j) = s ;
        }
    return (detW > 0)? detW * detW : -1. ;
}
double Pdfg::chlsk(double *w,double *q,int n,int *pNrang)
/* Cholesky decomposition of a definite, positive,
   symmetric matrix
   cf. L. Lebart et al.,
   Traitement des donnees statistiques,
   Dunod, pp 424, 1982.
   return = determinant of w or -1

   double *w ;      the input matrix
   double *q ;      the matrice in a triangle form
   int n ;          it is a n X n matrix
   int *pNrang ;    the  rank of w (must be  n )
*/
{
    int i, j, k ;
```

```
    register double *qjk , *qik, *qij ;
    double  *wij ;
    double s, det ;

    *pNrang = n ;
    det = 1. ;
    qij = &Q(0,0) ;
    wij = &W(0,0) ;
    for (i = 0 ; i < n ; i++)
        for (j = 0 ; j <= i ; j++, qij++, wij++) {
            s = *wij ;
            qjk = &Q(j,0) ;
            qik = &Q(i,0) ;
            for (k = 0 ; k < j ; k++)
                s -= *qjk++ * *qik++ ;
            if (i == j)
                if (fabs(s / *wij ) < EPSI) {
                    *qij = 0. ;
                    (*pNrang)-- ;
                    det = 0. ;
                }
                else
                    if (s < 0.)
                        return -1. ;
// non definite, positive
                    else {
                        *qij = sqrt(s) ;
                        det *= *qij ;
                    }
            else /* i != j */
                if (Q(j,j) == 0.)
                    *qij = 0 ;
                else
                    *qij = s / Q(j,j) ;
        } // for each column j
    return det ;
}
ostream & operator << (ostream & s, const Pdfg & item)
////////////////////////////////////////////////////
// prints a gaussian pdf
{
    s << endl ;
```

```
    s << "mean: " ;
    for (int j = 0 ; j < item.nCoef ; j++)
        s << item.mean[j] << " " ;
    s << endl ;
    s << "inv correl.: " << endl ;
    double* pf1 = item.covar ;
    for (int i = 0 ; i < item.nCoef ; i++) {
        for (int j = 0 ; j <= i ; j++, pf1++)
            s << *pf1 << " " ;
        s << endl ;
    }
    return s ;
}
```

2.   THE CENTROID CLASS

This class represents a cluster during the clustering process. It uses objects like Pdfg. Basically, a centroid is made up with two Gaussian *pdf*. One *pdf* oldg represents the current feature whereas newg is being estimated. This class is handled by the following methods:

addSample(x) adds the sample x to newg;

distance(x) computes the Mahalanobis distance between x and this current cluster ;

normCentro() terminates the estimation process, inverts the covariance matrix and swap the *pdf* in the cluster.

The file centroide.h gives the interface of the class Centroid.

```
#ifndef __CENTRO_H
#define __CENTRO_H
#include "Pdfg.h"
extern double MahaCovar[NCOEF * (NCOEF +1) / 2] ;
class  Centroide  // define a cluster
{
 private:
    Pdfg *oldg ;
    Pdfg *newg ;
    int nbr ;       // new nmb. of elements
    int oldNbr ;    // old nbr.

 public:
    Centroide *next ;   // link to next centroid
```

```
       Centroide() ;
       ~Centroide() ;
       Centroide(const Centroide&) ;
       Centroide(Centroide* p, double f) ;
       int effectif() const {return oldNbr ;}
       double distance(double *samn) const ;
       // distance between samn and class
       void addSample(const double *samn) ;
       // add sample to class
       double normCentro() ; // normalize
       double *getMat() const {return oldg -> getCovar() ;}
       friend ostream& operator << (ostream & s,
                                    const Centroide & item) ;
} ;
#endif // !__CENTRO_H
```

The file `centroide.C` gives the implementation of the class `Centroid`.

```
#include <math.h>
#include "centroide.h"
Centroide::Centroide()
/////////////////////////
{
    nbr = oldNbr = 0 ;
    next = NULL ;
    oldg = new Pdfg ;
    newg = new Pdfg ;
}
Centroide::~Centroide()
/////////////////////////
{
    delete oldg ;
    delete newg ;
    // do  not delete the following in the list
}
Centroide::Centroide(const Centroide& cg)
    : nbr (cg.nbr),
      oldNbr (cg.oldNbr),
      next (NULL)
{
    oldg = new Pdfg(*(cg.oldg)) ;
    newg = new Pdfg(*(cg.newg)) ;
}
```

```
Centroide::Centroide(Centroide* p, double f)
// clone and perturb both
    : nbr (0),
      oldNbr (p -> oldNbr),
      next (NULL)
{
    oldg = new Pdfg(*(p -> oldg)) ;
    oldg -> shake(f) ; // apply a f % factor to shake
    newg = new Pdfg() ;
    p -> oldg -> shake(-f) ; // shake p
}
double Centroide::distance(double* samn) const
// compute distance between samn and class
// represented by oldg
// using the Mahalanobis distance
{
    const double *cmean = oldg -> getMean() ;
    double xm[NCOEF] ;
    for (int i = 0 ; i < NCOEF ; i++)
        xm[i] = samn[i] - cmean[i] ;

    double sum1 = 0. ;
    double sum2 = 0. ;
    double *pf = MahaCovar ;
    for (int i = 0 ; i < NCOEF ; i++) {
        for( int j = 0 ; j < i ; j++)
            sum2 += xm[i] * xm[j] * *pf++ ;
        sum1 += *pf++ * xm[i] * xm[i] ;
    }
    return sum1 + 2. * sum2 ;
}
void Centroide::addSample(const double *samn)
{
    newg -> addWeightedSample(1., samn) ;
    nbr++ ;
}
double Centroide::normCentro()
{
    double det = newg -> normMat(nbr) ;
    if (det <= 0.) oldNbr = 0 ;
    else oldNbr = nbr ;
    Pdfg* tmp = oldg ; // swap old and new centroid
```

```
        oldg = newg ;
        newg = tmp ;
        nbr = 0 ;
        newg -> resetMC() ;
        return det ;
}
ostream& operator << (ostream & s, const Centroide & item)
{
        s << "nbr = " << item.oldNbr << endl ;
        s << *(item.oldg) << endl ;
        return s ;
}
```

3.   THE TOP DOWN CLUSTERING PROGRAM

This program, known as the LBG clustering method, implements the clustering methods first described by Linde, Buzo and Gray [Linde et al., 1980, Gray, 1984]. It makes use of the centroid class and is written in C++. As a distance between clusters, we use the Mahalanobis distance computed on all the samples. We got the best results with a diagonal matrix.

```
#include "centroide.h"
...
#define RG(i,j) ((i) * (i+1)/2 + (j))
// index in a symmetric matrix
double MahaCovar[NCOEF * (NCOEF + 1) / 2] ;
...
int  main(int argc, char **argv)
{
        double samples[MAXTOTALFRAME][NCOEF] ;
        double det ;
// read input data into samples[]
// and set nSamples
        ...
            cout << "Use " << nSamples << " samples\n" ;
            Pdfg::setNCoef(NCOEF) ;
// set the class variables
// general centroid
            Centroid* tete = new Centroid ;
            Centroid *p, *ptr ;
//   compute Mean and Covar  of input frames
```

```
for (int j = 0 ;j < nSamples ; j++)
    tete -> addSample(&samples[j][0]) ;
det = tete -> normCentro() ;
if (det <= 0.) {
    cerr << "cannot invert at init time\n" ;
    exit(1) ;
}
// use diagonal Mahalanobis distance
memcpy((char *)(MahaCovar),
       (char *)(tete -> getMat()),
       sizeof(MahaCovar)) ;
for (int i = 0 ; i < NCOEF ; i++)
    for (int j = 0 ; j < i ; j++)
        MahaCovar[RG(i,j)] = 0. ;
cout << *tete ;
int nbClasse = 1 ;
int nbMinPts = 0 ;
while (nbClasse < nbClasseMax) {
    n = 0;
    nbClasse *= 2 ;
    //   random perturbation
    p = tete;
    while (p != NULL) {
        Centroid *pm = new Centroid (p, 0.20) ;
        // clone and perturb
        pm -> next =  p -> next; // link pm after p
        p -> next =  pm;
        p =  pm -> next;
    }
    //   optimal partition and re-affectation
    double distortion, oldDistortion =  1.0e+34 ;
    do {
        oldDistortion = distortion ;
        distortion =  0. ;
        for (int j = 0 ; j < nSamples ; j++){
            double distMin =  1.0e+34;
            for (p = tete; p != NULL; p = p->next) {
                dint = p -> distance(&samples[j][0]);
                if (dint < distMin) {// look for min
                    ptr =  p;
                    distMin =  dint; // current min
                }
```

```
                } // for each class
                distortion += distMin;
                ptr -> addSample(&samples[j][0]) ;
            } // for each sample
            // compute mean distortion
            distortion /= nSamples ;
            for (p = tete; p != NULL ; p = p->next){
                det = p -> normCentro() ;
                if (det <= 0.) {
                    cerr << "cannot invert" << endl ;
                }
            } /* for each class */
            cout << nbClasse << ": "
                << distortion << " " << ++n << endl ;
        } // until the optimal partition has been found
        while (fabs((oldDistortion - distortion) /
                (distortion + 1.)) >= 2.e-4) ;
// remove empty class
        p = tete ;
        Centroid* nTete = NULL ;
        while (p != NULL) {
            Centroid *pS = p -> next ;
            if (p -> effectif() > nbMinPts) {
                p -> next = nTete ;
                nTete  = p ;
            }
            else delete(p) ;
            p = pS ;
        }
        tete =  nTete ;
// print stat on current clustering
        p = tete ;
        while (p != NULL)  {
            cout << *p ;
            p = p -> next ;
        }
        cout << endl ;
    }
    return 0 ;
}
```

References

[Abramowitz and Stegun, 1965] Abramowitz, M. and Stegun, I. (1965). *Handbook of Mathematical Functions With Formulas, Graphs, and Mathematical Tables*. Dover.

[Akaike, 1974] Akaike, H. (1974). A New Look At Statistical Model Identification. *IEEE Trans. on Automatic Control*, 19:716 – 723.

[Arnold, 1994] Arnold, A. (1994). *Finite Transition Systems*. Prentice Hall.

[Arnold and Nivat, 1982] Arnold, A. and Nivat, M. (1982). Comportements de Processus. Technical report, LITP, Paris, France.

[Aycard, 1998] Aycard, O. (1998). *Architecture de Contrôle pour Robot Mobile en Environnement Intérieur Structuré*. PhD thesis, Université Henri Poincaré-Nancy 1.

[Aycard et al., 1997a] Aycard, O., Charpillet, F., Fohr, D., and Mari, J.-F. (1997a). Place Learning and Recognition Using Hidden Markov Models. In *Proceedings IEEE-RSJ on International Conference on Intelligent Robots and Systems*, pages 1741 – 1746, Grenoble, France.

[Aycard et al., 1997b] Aycard, O., Charpillet, F., and Haton, J. (1997b). A New Approach to Design Fuzzy Controllers for Mobile Robots Navigation. In *IEEE International Symposium on Computational Intelligence in Robotics and Automation*.

[Aycard et al., 1998] Aycard, O., Mari, J.-F., and Charpillet, F. (1998). Second Order Hidden Markov Models for Places Recognition: New Results. In *proceedings of IEEE International Conference on Tools with Artificial Intelligence*.

221

[Baeza-Yates, 1989] Baeza-Yates, R. (1989). A Trivial Algorithm Whose Analysis is not: A Continuation. *BIT*, (29):378–394.

[Bahl et al., 1993] Bahl, L. R., Bellarda, J. R., de Souza, P. V., Gopalakrishnan, P. S., Nahamoo, D., and Picheny, M. A. (1993). Multonic Markov Word Models for Large Vocabulary Continuous Speech Recognition . *IEEE Transactions on Speech and Audio Processing*, 1(3):334–344.

[Baker, 1975] Baker, J. (1975). The Dragon system- An overview. *IEEE Trans. on Acoutics, Speech and Signal Processing*, 23(11):24 – 29.

[Baker, 1974] Baker, J. K. (1974). Stochastic Modeling for Automatic Speech Understanding. In Reddy, D., editor, *Speech Recognition*, pages 521 – 542. Academic Press, New York, New-York.

[Beauquier et al., 1987] Beauquier, J., Bérard, B., and Thimonier, L. (1987). On a Concurrency Measure. In *Proc. Second I.S.C.I.S., Istanbul, Turkey, 1987*, pages 211–225.

[Bellman, 1957] Bellman, R. (1957). *Dynamic Programming*. Princeton Univ. Press.

[Billingsley, 1968] Billingsley, P. (1968). *Convergence of Probability Measures*. Wiley.

[Boissonnat et al., 1992] Boissonnat, J., Devillers, O., Schott, R., Teillaud, M., and Yvinec, M. (1992). Applications of Random Sampling to On-line Algorithms in Computational Geometry. *Discrete and Comput. Geometry*, (8):51–71.

[Boissonnat and Dobrind, 1992] Boissonnat, J. and Dobrind, K. (1992). Randomized Construction of the Upper Envelope of Triangles in the Space. In *Proc. 4th Canad. Conf. Comput. Geom.*, pages 311–315.

[Boissonnat and Yvinec, 1995] Boissonnat, J. and Yvinec, M. (1995). *Géométrie Algorithmique*. Ediscience International.

[Boutilier and Dearden, 1994] Boutilier, C. and Dearden, R. (1994). Using Abstractions for Decision-theoretic Planning With Time Constraints. In *Proceedings of the National Conference on Artificial Intelligence*, pages 1016–1022.

[Bucklew, 1990] Bucklew, J. (1990). *Large Deviations Techniques in Decision, Simulation, and Estimation*. Wiley.

[Buzo et al., 1980] Buzo, A., Gray, A. H., Gray, R. M., and Markel, J. D. (1980). Speech Coding Based upon Vector Quantization. *IEEE Trans. on Acoustics Speech Signal Processing*, 28(5):562 – 574.

[Calliope, 1989] Calliope (1989). *La parole et son traitement automatique.* Masson.

[Cardin et al., 1993] Cardin, R., Normandin, Y., and Millien, E. (1993). Inter-Word Coarticulation Modeling and MMIE Training for Improved Connected Digit Recognition. In *Proceedings of IEEE-International Conference On Acoustics, Speech, and Signal Processing*, volume 2, pages 243 – 246.

[Cassandra et al.,] Cassandra, A., Kaelbling, L., and Kurien, J. Acting under uncertainty: Discrete bayesian models for mobile-robot navigation. In *Proceedings of IEEE International Conference on Intelligent Robots and Systems*, pages 963–972.

[Castillo, 1988] Castillo, E. (1988). *Extreme Value Theory in Engineering.* Academic Press.

[Cerf, 1994] Cerf, R. (1994). *Une Théorie Asymptotique des Algorithmes Génétiques.* PhD thesis, Université Monpellier II.

[Chauvin et al., 1999] Chauvin, B., Drmota, M., and Jabbour-Hattab, J. (1999). The Profile of Binary Search Trees. Technical report, Université de Versailles, France.

[Chollet and Gagnoulet, 1982] Chollet, G. and Gagnoulet, C. (1982). On the Evaluation of Speech and Data Bases Using a Reference System. In *Proceedings of IEEE-International Conference On Acoustics, Speech, and Signal Processing*, volume 3, pages 2026 – 2029, Paris.

[Chung, 1976] Chung, K. (1976). Excursions in Brownian Motion. *Ark. Math.*, (14):155–177.

[Chung and Williams, 1983] Chung, K. and Williams, R. (1983). *Introduction to Stochastic Integration.* Birkhäuser.

[Clarkson et al., 1993] Clarkson, K., Mehlhorn, K., and Seidel, R. (1993). Four Results on Randomized Incremental Constructions. *Comput. Geom. Theory Appl.*, 3(4):185–212.

[Clarkson and Shor, 1989] Clarkson, K. and Shor, P. (1989). Applications of Random Sampling in Computational Geometry. *Discrete Comput. Geom*, (4):387–421.

[Cole et al., 1992] Cole, R., Roginski, K., and Fanty, M. (1992). A Telephone Speech Database of Spelled and Spoken Name. In *Proceedings of International Conference on Spoken Language Processing*, pages 891–895, Banff.

[Cox and Miller, 1965] Cox, D. and Miller, H. (1965). *The Theory of Stochastic Processes*. Chapman and Hall.

[Crystal and House, 1988] Crystal, T. and House, A. (1988). Segmental Durations in Connected Speech Signals: Current Results. *Journal of Acoustical Society of America*, 83(4):1553 – 1573.

[Csaki et al., 1987] Csaki, E., Földes, A., and Salminen, P. (1987). On the Joint Distribution of the Maximum and its Location for a Linear Diffusion. *Ann. Inst. H. Poincaré Probab. Statist.*, (23):179–194.

[Daniels, 1989] Daniels, H. (1989). The Maximum of a Gaussian Process Whose Mean Path has a Maximum, with an Application to the Strength of Bundles of Fibres. *Adv. Appl. Probab.*, (21):315–333.

[Daniels and Skyrme, 1985] Daniels, H. and Skyrme, T. (1985). The Maximum of a Random Walk Whose Mean Path has a Maximum. *Adv. Appl. Probab.*, (17):85–99.

[Davis and Mermelstein, 1980] Davis, S. and Mermelstein, P. (1980). Comparaison of Parametric Representations for Monosyllabic Word Recognition in Continuously Spoken Sentences . *IEEE Transactions on Acoutics, Speech and Signal Processing*, 28(4):357 – 366.

[Dean et al.,] Dean, T., Kaelbling, L., Kirman, J., and Nicholson, A. Planning with deadlines in stochastic domains. In *Proceedings of the 11th National Conference on Artificial Intelligence*, pages 574–579.

[Dean and Lin, 1995] Dean, T. and Lin, S. (1995). Decomposition Techniques for Planning in Stochastic Domains. In *Proceedings of International Joint Conference on Artificial Intelligence*.

[Dembo and Zeitouni, 1993] Dembo, A. and Zeitouni, O. (1993). *Large Deviations and Applications*. Jones and Barlet.

[Dempster et al., 1977] Dempster, A., Laird, N., and Rubin, D. (1977). Maximum-Likelihood From Incomplete Data Via The EM Algorithm. *Journal of Royal Statistic Society, Ser. B (methodological)*, 39:1 – 38.

[Devillers, 1996] Devillers, O. (1996). An Introduction to Randomization in Computational Geometry. *Theoretical Comput. Science*, 157:35–52.

[Devillers et al., 1992] Devillers, O., Meiser, S., and Teillaud, M. (1992). Fully Dynamic Delaunay Triangulation in Logarithmic Expected Time per Operation. *Comp. Geom. Theor. Appl.*, 2(2):55–80.

[Devroye, 1986] Devroye, L. (1986). A Note on the Height of Binary Search Trees. *J. Assoc. Comput. Mach.*, (33):489–498.

[Devroye, 1987] Devroye, L. (1987). Branching Processes in the Analysis of the Heights of Trees. *Acta Inform.*, (24):277–298.

[Devroye, 1992] Devroye, L. (1992). A Limit Theorem for Random Skip Lists. *The Annals of Applied Probability*, 2(3):597–609.

[Diday et al., 1982] Diday, E., Lemaire, J., Pouget, J., and Testu, F. (1982). *Éléments d'analyse de données*. Dunod.

[du Preez, 1998] du Preez, J. A. (1998). *Efficient High-Order Hidden Markov Modelling*. PhD thesis, University of Stellenbosh.

[Durbin, 1985] Durbin, J. (1985). The First-passage Density of a Continuous Gaussian Process to a General Boundary. *J. Appl. Probab.*, (22):99–122.

[Elfes, 1989] Elfes, A. (1989). Using Occupancy Grids for Mobile Robot Perception and Navigation. *IEEE Computer*, 22(6):46–57.

[Ellis, 1977] Ellis, C. A. (1977). Probabilistic Model of Computer Deadlock. *IEEE Trans. on ASSP*, (12):43 – 60.

[Feller, 1970] Feller, W. (1970). *An Introduction To Probability Theory and Its Applications*, volume 2 volumes. Wiley.

[Flajolet, 1986] Flajolet, P. (1986). The Evolution of Two Stacks in Bounded Space and Random Walks in a Triangle. volume 233 of *Lecture Notes in Computer Science*, pages 325–340.

[Flajolet, 1987] Flajolet, P. (1987). Analytic Models and Ambiguity of Context-free Languages. *Theoret. Comput. Sci*, (49):283– 309.

[Flajolet et al., 1981] Flajolet, P., Françon, J., and Vuillemin, J. (1981). Sequence of Operations Analysis for Dynamic Data Structures. *J. of Algorithms*, (1):111– 141.

[Forney, 1973] Forney, G. (1973). The Viterbi Algorithm. *IEEE Transactions*, 61:268–278.

[Françon, 1978] Françon, J. (1978). Histoires de fichiers. *RAIRO Inf. Th.*, (12):49– 62.

[Françon, 1986] Françon, J. (1986). A Quantitative Approach of Mutual Exclusion. *Informatique Théorique et Applications/Theoretical Informatics and Applications*, (20):275–289.

[Françon et al., 1990] Françon, J., Randrianarimanana, B., and Schott, R. (1990). Analysis of Dynamic Algorithms in D.E. Knuth's Model. *Theoretical Computer Science*, (72):147– 167.

[Freidlin and Wentzell, 1984] Freidlin, M. and Wentzell, A. (1984). *Random Perturbations of Dynamical Systems*. Springer-Verlag.

[Furui, 1986] Furui, S. (1986). Speaker-independent Isolated Word recognition Using Dynamic Features of Speech Spectrum. *IEEE Transactions on Acoutics, Speech and Signal Processing.*

[Furui, 1981] Furui, S. (April 1981). Cepstrum Analysis Techniques for Automatic Speaker Verification. *IEEE Transactions on Acoutics, Speech and Signal Processing.*

[Garofolo et al., 1993] Garofolo, J., Lamel, L., Fisher, W., Fiscus, J., Pallet, D., and Dahlgren, N. (1993). The DARPA TIMIT Acoustic-Phonetic Continuous Speech Corpus CDROM .

[Geniet et al., 1996] Geniet, D., Schott, R., and Thimonier, L. (1996). A Markovian Concurrency Measure. *Informatique Théorique et Applications/Theoretical Informatics and Applications*, 30(4):295– 304.

[Goldberg, 1989] Goldberg, D. (1989). *Genetic Algorithms in Search, Optimization and Machine Learning*. Addison-Wesley.

[Gordon, 1999] Gordon, A. (1999). *Classification 2nd Edition*. Chapman & Hall/CRC.

[Gray, 1984] Gray, R. M. (1984). Vector Quantization. *IEEE ASSP Magazine*, 1(2):4 – 29.

[Haberman, 1978] Haberman, A. N. (1978). *System Deadlocks*, volume 3, chapter 7, pages 256–297. Prentice-Hall.

[Haeb-Umbach et al., 1993] Haeb-Umbach, R., Geller, D., and Ney, H. (1993). Improvements in Connected Digit Recognition Using Linear Discriminant Analysis and Mixture Densities. In *Proceedings of IEEE-International Conference On Acoustics, Speech, and Signal Processing*, pages 239 – 242.

[Hanson and Applebaum, 1990] Hanson, B. A. and Applebaum, T. (1990). Robust Speaker-Independent Word Recognition Using Static,

Dynamic, and Acceleration Features: Experiments with Lombard and Noisy Speech. In *Proceedings of IEEE-International Conference On Acoustics, Speech, and Signal Processing*, pages 857 – 860.

[Hauskrecht et al., 1998] Hauskrecht, M., Meuleau, N., Kaelbling, L., Dean, T., and Boutilier, C. (1998). Hierarchical Solution of Markov Decision Processes Using Macro-actions. In *Proceedings of the Fourteenth Conference on Uncertainty in Artificial Intelligence*.

[Hida, 1980] Hida, T. (1980). *Brownian Motion*. Springer Verlag.

[Holland, 1975] Holland, J. (1975). *Adaptation in Natural and Artificial Systems*. The University of Michigan Press, Ann Arbor.

[Howard, 1960] Howard, R. (1960). *Dynamic Programming and Markov Processes*. MIT Press, Cambridge, Massachussets.

[Ito and McKean, 1974] Ito, K. and McKean, H. (1974). *Diffusion Processes and Their Sample Paths*, volume 3. Springer-Verlag.

[Jelinek, 1976] Jelinek, F. (1976). Continuous Speech Recognition by Statistical Methods. *IEEE Trans. on Acoutics, Speech and Signal Processing*, 64(4):532 – 556.

[Jonassen and Knuth, 1978] Jonassen, A. and Knuth, D. E. (1978). A Trivial Algorithm Whose Analysis Isn't. *Journal of Computing System Sci.*, (16):301 – 332.

[Junqua et al., 1995a] Junqua, J., Valente, S., Fohr, D., and Mari, J.-F. (1995a). An N-Best Strategy, Dynamic Grammars and Selectively Trained Neural Networks for Real-Time Recognition of Continuously Spelled Names over the Telephone. In *Proc. ICASSP*, pages 852 – 855, Detroit.

[Junqua et al., 1995b] Junqua, J.-C., Fohr, D., Mari, J.-F., Applebaum, T. H., and Hanson, B. H. (1995b). Time derivatives, Cepstral Normalization, and Parameter Filtering for Continuously Spelled Names over the Telephone. In *Proceedings 4th European Conference on Speech Communication and Technology*, volume 1, pages 1385–1388, Madrid (Spain).

[Kenyon-Mathieu and Vitter, 1991] Kenyon-Mathieu, C. and Vitter, J. S. (1991). The Maximum Size of Dynamic Data Structures. *SIAM J. Comput.*, (20):807–823.

[Knessl et al., 1985] Knessl, C., Matkowsky, B. J., Schuss, Z., and Tier, C. (1985). An Asymptotic Theory of Large Deviations for Markov Jump Processes. *SIAM J. Appl. Math*, (46):1006–1028.

[Knott, 1975] Knott, G. (1975). *Deletion in Binary Storage Trees*. PhD thesis, Stanford University.

[Knuth, 1973] Knuth, D. E. (1973). *The Art of Computer Programming*, volume 1. Addison-Wesley.

[Knuth, 1977] Knuth, D. E. (1977). Deletions That Preserve Randomness. *Trans. Software Eng.*, 3:351 – 359.

[Koenig and Simmons, 1998] Koenig, S. and Simmons, R. (1998). Xavier: A Robot Navigation Architecture Based on POMDP Models. In *Artificial Intelligence Based Mobile Robotics*. MIT Press.

[Kortenkamp and Weymouth, 1994] Kortenkamp, D. and Weymouth, T. (1994). Topological Mapping for Mobile Robots Using a Combination of Sonar and Vision Sensing. In *Proceedings of AAAI*.

[Kriouile, 1990] Kriouile, A. (1990). *La reconnaissance automatique de la parole et les modèles de Markov cachés : modèles du second ordre et distatnce de Viterbi à optimalité locale*. PhD thesis, Université de NANCY 1.

[Lamel and Gauvain, 1993] Lamel, L. and Gauvain, J. (1993). High Performance Speaker-Independent Phone Recognition using CDHMM. In *Proceedings of European Conference on Speech Communication and Technology*, pages 121 – 124.

[Laroche and Charpillet, 1998] Laroche, P. and Charpillet, F. (1998). State Aggregation for Solving Markov Decision Problems - An Application to Mobile Robotics. In *IEEE International Conference Tools with Artificial Intelligence*.

[Laroche et al., 1999] Laroche, P., Charpillet, J., and Schott, R. (1999). Mobile Robotics Planning Using Abstract Markov Decision Processes. pages 299–306. ICTAI'99, IEEE.

[Lebart et al., 1982] Lebart, L., Morineau, A., and Fenelon, J. (1982). *Traitement des données statistiques : Méthodes et programmes*. Dunod.

[Lebart et al., 1995] Lebart, L., Morineau, A., and Pinon, M. (1995). *Statistique exploratoire multidimensionnelle*. Dunod.

[Lee and Hon, 1989] Lee, K.-F. and Hon, H.-W. (1989). Speaker-Independent Phone Recognition Using Hidden Markov Models. *IEEE Transactions on Acoutics, Speech and Signal Processing*, 37(11):1641 – 1648.

[Levinson, 1986] Levinson, S. E. (1986). Continuously Variable Duration Hidden Markov Models for Automatic Speech Recognition. *Computer Speech and Language*, 1:29 – 45.

[Linde et al., 1980] Linde, Y., Buzo, A., and Gray, R.-M. (1980). An Algorithm for Vector Quantizer Design. *IEEE Trans. on Communications*, com-28(1):84 – 94.

[Liporace, 1982] Liporace, L. A. (1982). Maximum Likelihood Estimation for Multivariate Observations of Markov Sources. *IEEE Transactions on Information Theory*, 28(5):729 – 734.

[Littman et al.,] Littman, M., Dean, T., and Kaelbling, L. On the complexity of solving markov decision problems. *Proceedings of the Eleventh Annual Conference on Uncertainty in Artificial Intelligence (UAI-95), Montreal, Québec, Canada*, pages 394–402.

[Louchard, 1987] Louchard, G. (1987). Random Walks, Gaussian Processes and List Structures. *Theoretical Computer Science*, (53):99–124.

[Louchard et al., 1997] Louchard, G., Kenyon, C., and Schott, R. (1997). Data Structures Maxima. *SIAM J. Comput.*, (4):1006–1042.

[Louchard et al., 1992] Louchard, G., Randrianarimanana, B., and Schott, R. (1992). Probabilistic Analysis of Dynamic Algorithm in D.E. Knuth's Model. *Theoretical Computer Science*, (93):201– 225.

[Louchard and Schott, 1991] Louchard, G. and Schott, R. (1991). Probabilistic Analysis of Some Distributed Algorithms. *Random Structures and Algorithms*, (2):151–186.

[Louchard et al., 1994] Louchard, G., Schott, R., Tolley, M., and Zimmermann, P. (1994). Random Walks, Heat Equations and Distributed Algorithms. *J. Comp. Appl. Math.*, (53):243–274.

[Mahmoud, 1992] Mahmoud, H. (1992). *Evolution of Random Search Trees*. J. Wiley & Sons.

[Maier, 1991a] Maier, R. (1991a). A path integral approach to data structures evolution. *Journal of Complexity*, 7(3):232– 260.

[Maier, 1991b] Maier, R. S. (1991b). Colliding Stacks: A Large Deviations Analysis. *Random Structures and Algorithm*, (2):379–420.

[Maier and O'Cinneide, 1992] Maier, R. S. and O'Cinneide, C. A.
(1992). A Closure Characterization of Phase-Type Distributions.
J. Appl. Probab., (29):92–103.

[Maier and Schott, 1993a] Maier, R. S. and Schott, R. (1993a). Exhaustion of Shared Memory: Stochastic Results. volume 709 of *LNCS*,
pages 494–505. WADS'93, Springer Verlag.

[Maier and Schott, 1993b] Maier, R. S. and Schott, R. (1993b). Regular
Approximations to Shuffle Products of Context-free Languages, and
Convergence of Their Generating Functions. volume 710 of *LNCS*,
pages 352–362. FCT'93, Springer Verlag.

[Mari et al., 1997] Mari, J.-F., Haton, J.-P., and Kriouile, A. (1997).
Automatic Word Recognition Based on Second-Order Hidden Markov
Models. *IEEE Transactions on Speech and Audio Processing*, 5:22 –
25.

[McDiarmid, 1996] McDiarmid, C. (1996). Large Deviations for Quicksort. *J. of Algorithms*, (21):476–507.

[Minc, 1988] Minc, X. (1988). *Nonnegative Matrices*, volume 1. Wiley.

[Morignot et al., 1997] Morignot, P., Aycard, O., and Charpillet, F.
(1997). A Pair of Heterogeneous Agent in a Unique Vehicle for Object
Motion. In *Proceeding of the IEEE/ICTAI*, pages 508 – 513.

[Motvani and Raghavan, 1995] Motvani, R. and Raghavan, P. (1995).
Randomized Algorithms. Cambridge University Press.

[Mulmuley, 1993] Mulmuley, K. (1993). *Computational Geometry: An
Introduction Through Randomized Algorithms*. Prentice Hall.

[Naeh et al., 1990] Naeh, T., Klosek, M. M., Matkowsky, B. J., and
Schuss, Z. (1990). A Direct Approach to the Exit Problem. *SIAM
J. Appl. Math.*, (50):595–627.

[Nomadics, 1996] Nomadics (1996). Nomad 200 User's Manual. Technical report, Nomadics.

[Normandin et al., 1994] Normandin, Y., Cardinand, R., and Mori,
R. D. (1994). High-Performance Connected Digit Recognition Using Maximum Mutual Information Estimation . *IEEE Transactions
on Speech and Audio Processing*, 2(2):299–311.

[P. W. Purdom and Stigler, 1970] P. W. Purdom, J. and Stigler, S. M. (1970). Statistical Properties of the Buddy System. *J. Assoc. Comput. Mach.*, (17):683–697.

[Papadakis et al., 1992] Papadakis, T., Munro, J., and Poblete, P. (1992). Average Search and Update Costs in Skip Lists. *BIT*, (32):316–332.

[Papadimitriou and Tsitsiklis, 1987] Papadimitriou, C. and Tsitsiklis, J. (1987). The Complexity of Markov Decision Processes. *Mathematics of Operations Research*, 12(3):441–450.

[Parr, 1998] Parr, R. (1998). Flexible Decomposition Algorithms for Weakly Coupled Markov Decision Problems. In *Proceedings of the Fourteenth Conference on Uncertainty in Artificial Intelligence.*

[Precup and Sutton, 1997] Precup, D. and Sutton, R. S. (1997). Multi-time models for temporally abstract planning. In *Advances in Neural Information Processing Systems.* MIT Press.

[Pugh, 1990] Pugh, W. (1990). Skip Lists: A Probabilistic Alternative to Balanced Trees. *Communications of the ACM*, 33(6):668–676.

[Puterman, 1994] Puterman, M. (1994). *Markov Decision Processes.* John Wiley & Sons, New York.

[Rabiner and Juang, 1995] Rabiner, L. and Juang, B. (1995). *Fundamentals of Speech Recognition.* Prentice Hall.

[Rathinavelu and Deng, 1995] Rathinavelu, C. and Deng, L. (1995). Use of Generalized Dynamic Feature Parameters for Speech Recognition: Maximum Likelihood and Minimum Classifcation Error Approaches. In *Proceedings International Conference on Spoken Language Processing*, pages 373 – 376, Detroit.

[Redner and Walker, 1984] Redner, R. and Walker, H. (1984). Mixture Densities, Maximum Likelihood and The EM Algorithm. *SIAM*, 26(2):195 – 239.

[Régnier, 1989] Régnier, M. (1989). A Limiting Theorem for Quicksort. *Informatique Théorique et Applications/Theoretical Informatics and Applications*, pages 335–343.

[Revuz, 1975] Revuz, D. (1975). *Markov Chains.* North-Holland.

[Rhee and Talagrand, 1987] Rhee, W. and Talagrand, M. (1987). Martingale Inequalities and NP-complete Problems. *Math. Oper. Research*, 12(1):177–181.

[Rissanen, 1978] Rissanen, J. (1978). Modeling By Shortest Data Description. *Automatica*, 14:465 – 471.

[Rompais et al., 1995] Rompais, C., Barret, J.-F., and Pradet, G. (1995). Neural Classifier for Environment Recognition in Mobile Robotics . In *Proceeding of the IEEE/ICANN*, pages 463 – 468.

[Rösler, 1989] Rösler, U. (1989). A Limit Theorem for "Quicksort". *Informatique Théorique et Applications/Theoretical Informatics and Applications*, 25(1):85–100.

[Russell and Cook, 1887] Russell, M. J. and Cook, A. (1887). Experimental Evaluation of Duration Modelling Techniques For Automatic Speech Recognition. In *Proceedings of IEEE-International Conference On Acoustics, Speech, and Signal Processing*, pages 2376 – 2379, Dallas.

[Sakoe and Chiba, 1978] Sakoe, H. and Chiba, S. (1978). Dynamic Programming Optimization for Spoken Word Recognition, . *IEEE Trans. on Acoutics, Speech and Signal Processing*, 26(11):43 – 49.

[Salminen, 1988] Salminen, P. (1988). On the First Hitting Time and the Last Exit Time for a Brownian To/From a Moving Boundary. *Adv. Appl. Probab.*, (20):411–426.

[Salomaa and Soittola, 1978] Salomaa, A. and Soittola, M. (1978). *Automata-Theoretic Aspects of Formal Power Series*, volume 1. Springer Verlag.

[Saporta, 1990] Saporta, G. (1990). *Théories et méthodes de la statistique*. Publications de l'institut français du pétrole.

[Schwartz and Austin, 1991] Schwartz, R. and Austin, S. (1991). A Comparison of Several Approximate Algorithms for Finding Multiple (N-BEST) Sentence Hypotheses. In *Proceedings of IEEE-International Conference On Acoustics, Speech, and Signal Processing*, pages 701 – 704.

[Schwarz, 1978] Schwarz, G. (1978). Estimating the Dimension Of a Model. *Ann. Stat.*, 6(2):461 – 464.

[Sedgewick, 1985] Sedgewick, B. (1985). *Algorithms*. Addison Wesley.

[Shibata, 1986] Shibata, R. (1986). Criteria Of Statistical Model Selection. Technical report, Dpt. of Math., Fac. of Science and Techn.

[Simmons and Koenig, 1995] Simmons, R. and Koenig, S. (1995). Probabilistic robot navigation in partially observable environments. In *Proceedings of International Joint Conference on Artificial Intelligence*, pages 1080–1087.

[Stroock, 1984] Stroock, D. (1984). *An Introduction to the Theory of Large Deviations*. Springer Verlag.

[Suaudeau and André-Obrecht, 1993] Suaudeau, N. and André-Obrecht, R. (1993). Sound Duration Modelling and time variable Speaking rate in a Speech Recognition System. In *Proceedings of European Conference on Speech Communication and Technology*, pages 307 – 310.

[Tou and Gonzales, 1974] Tou, J. T. and Gonzales, R. (1974). *Pattern Recognition Principles*. Addison-Wesley.

[Varadhan, 1984] Varadhan, S. (1984). *Large Deviations and Applications*. SIAM.

[Williams and BairdIII, 1993] Williams, R. and BairdIII, L. (1993). Tight performance bounds on greedy policies based on imperfect value functions. Technical Report NU-CCS-93-14, Northeastern University.

[Wilpon et al., 1993] Wilpon, J. G., Lee, C.-H., and Rabiner, L. R. (1993). Connected Digit Recognition Based on Improved Acoustic Resolution. *Computer Speech and Language*, 7:15 – 26.

[Wilpon and Rabiner, 1985] Wilpon, J. G. and Rabiner, L. R. (1985). A Modified K-Means Clustering Algorithm for Use in Isolated Work Recognition. *IEEE Trans. ASSP*, 33(3):587 – 594.

[Wong, 1982] Wong, M. A. (1982). A Hybrid Clustering Method for Identifying High Density Clusters. *Journal of Amer. Statist. Assoc.*

[Wyk and Vitter, 1986] Wyk, C. V. and Vitter, J. (1986). The Complexity of Hashing With Lazy Deletions. *Algorithmica*, (1):17–29.

[Yamauchi, 1995] Yamauchi, B. (1995). *Exploration and Spatial Learning in Dynamic Environnements*. PhD thesis, Case Western Reserve University.

[Yamauchi and Langley, 1996] Yamauchi, B. and Langley, P. (1996). Place Learning in Dynamic Real-World Environments. In *Proceedings of ROBOLEARN-96: International Workshop for Learning in Autonomous Robots*, pages 123 – 129.

[Yao, 1981] Yao, A. C. (1981). An Analysis of a Memory Allocation Scheme for Implementing Stacks. *SIAM J. Comput*, (10):398–403.

Index

boundary, 116

absorbing barrier, 116
aggregation, 200
aggregation criteria, 48
AIC, 55
akaike, 55
Akaike criterion, 55
allocation algorithm, 115
Arnold-Nivat's model, 107
array, 70
asymptotic convergence, 101
automata, 107
average case analysis, 60

banker algorithm, 120
Bayes formula, 41
Bernoulli random variable, 7
BIC, 55
Bieynamé-Tchebychev inequality, 8
binary search, 60
binomial random variable, 7
broken corner, 120
Brownian bridge, 89
Brownian excursion, 73
Brownian motion, 28, 89, 90, 94, 117, 118

central limit theorem, 25, 31
characteristic function, 25, 39
Chernoff inequality, 8
chisquare test, 32
chromosome, 99
classification, 41
cluster, 44, 205
clustering, 41
concurrency measure, 107
concurrent systems, 108
conditional mean, 39
constraint, 74
correlation coefficient, 24

correlation matrix, 44
covariance, 23, 35
critical section, 112
crossover, 99

Daniels's fundamental result, 89
data structure, 63
data structure's maxima, 88
data type, 69
deadlock, 107
decomposition, 197
deletion, 63
dictionary, 69
die, 3, 33
dispersion, 34
distance, 34, 44
distortion, 50
distributed algorithm, 115
dynamic algorithm, 69

EM algorithm, 35
environment, 193
execution, 111
exponential random variable, 21

fair, 33
free monoid, 70, 107
fundamental matrix, 12

Gaussian distribution, 35
gaussian random variable, 21
genetic algorithm, 99
geometric random variable, 7

heap, 70
hidden Markov model, 35, 179
hierarchy, 48
history, 70
hitting place, 118
hitting time, 117

HMM2, 16

indoor navigation, 197
inertia, 45
information, 34
insertion, 63

Jensen inequality, 38

K-means, 50
Knuth's model, 71
Koenig-Huygens theorem, 47
Kullback-Leibler measure, 34, 55

law of large numbers, 9
limiting distribution, 63
limiting profile, 94
Lindeberg condition, 26
linear list, 69, 72, 80
linked list, 70

Mahalanobis distance, 35, 171
marginal distribution, 22
Markov chain, 10, 101
Markov decision processes, 19, 191
Markov inequality, 8
markovian model, 71, 72
martingale, 61
martingales, 20
maximum likelihood, 35, 38
moment, 7
morphism, 108
mutation, 99

navigation architecture, 195
negative query, 64

observation function, 196
optimal policy, 204
optimality criteria, 193
Ornstein-Uhlenbeck process, 29

pagoda, 70
parallel processing, 107
pattern recognition, 34
performance, 113
planning algorithm, 193, 198
Poisson random variable, 7
policy, 196
policy iteration, 193

POMDP, 20
positive query, 64
possibility, 3
possibility function, 70
priority queue, 69
probabilistic automaton, 108
probability distribution, 4
probability generating function, 9

quantitative data, 41
quicksort, 59

random walk, 115
rectangle, 120
reflecting barrier, 115
reflection principle, 75
regression line, 24
regular language, 107
request, 112
rissanen, 55
robotics, 191

sample, 34
schema, 70
selection, 99
shared storage, 115
sorting algorithm, 59
speech recognition, 36
stack, 69
state aggregation, 197
state estimator, 195
stationary distribution, 13
symbol table, 70
synchronized automaton, 111

test, 31
trace, 35
transition, 40
transition probabilities, 40
trend-free case, 117
triangle, 115
two stacks problem, 115

urn, 31

value iteration, 193
variance, 46
Viterbi algorithm, 14

Ward's algorithm, 50, 175